Public Address Systems

Public Address Systems

Vivian Capel

Focal Press

Focal
An imprint of Butterworth-Heinemann Ltd
Linacre House, Jordan Hill, Oxford OX2 8DP

A member of the Reed Elsevier plc group

OXFORD LONDON BOSTON
MUNICH NEW DELHI SINGAPORE SYDNEY
TOKYO TORONTO WELLINGTON

First published 1992
Paperback edition 1995

British Library Cataloguing in Publication Data
Capel, Vivian
Public Address Systems
I. Title
621.389

ISBN 0 240 51423 8

Library of Congress Cataloguing in Publication Data
A catalogue record for this book is available from the Library of Congress

Typeset by STM Typesetting Ltd, Amesbury, Wilts.
Printed and bound in Great Britain by Athenaeum Press Ltd, Gateshead, Tyne & Wear

Contents

Preface

Of all man's accomplishments, the field of communication is probably his most outstanding. Fibre-optics, direct intercontinental telephone dialling, mobile telephones, communications satellites, interplanetary probes all demonstrate his ability and fulfill his need to communicate with ease over great distances.

Most of these developments are fairly recent, but a much older yet no less important need is to communicate audibly with large numbers of his fellows at closer quarters, in the same location.

In earliest times this was done by relay speakers, who stood within earshot of the original speaker and then repeated what he said at suitable intervals. Others, further away, did likewise until an audience of thousands or even tens of thousands heard what had been said. Some knowledgeable speakers used natural topography or the carrying power of sound over water to reach large audiences.

Nowadays we use a public-address system, and while progress may not have been as spectacular as in some other areas of communication, there have been steady improvements and refinements so that the modern system is a far cry from the often inaudible, tinny, whistling, unwieldy arrangements of a few decades ago.

Here, we discuss basic requirements, audio engineering principles, microphones, mixers, amplifiers, loudspeakers, distribution systems, avoiding feedback, induction loops, outdoor systems, live music, achieving reliability, as well as many practical tips, test equipment, and common faults. In short almost everything the practical p.a. man needs to know. The object is to deal with these in a manner that will be understandable to readers with limited technical knowledge, yet also be of value to the professional public-address engineer.

Vivian Capel

1 Basic requirements

To design and install an effective public-address system we need to determine just what is needed, what we are aiming for and how it can be achieved. The first necessity is that the sound must be at a sufficiently high volume level with minimum variation to be comfortably heard at all parts of the hall by everyone with reasonable hearing. (The deaf can be catered for by an induction loop to be dealt with later.) This is the obvious requirement, but in a surprising number of installations it is not achieved. In others it is considered to be the only requirement, and volume is heard at the expense of clarity.

Equally important though is the second: intelligibility. This is harder to effect and is the one that so many systems fall down on. The purpose of a public address system is to relay speech, to provide the communication link between the speaker and his audience. Nothing is accomplished by a system that enables the audience to hear the sound of the speaker's voice without understanding what he is saying, yet many do just that.

A third requirement, which though less important, is highly desirable if it can be obtained without detriment to the other two, is naturalness. Ideally, an audience should not be aware that a public address system is in operation at all, they should just hear what appears to be the natural sound of the speaker's voice coming from his direction at comfortable volume.

A fourth necessity is reliability, especially with a larger installation. A breakdown during an important meeting could be a catastrophe.

Adequate sound level

Going back to that first requirement, volume is governed to a large extent by the great bugbear of all public address systems, feedback. Any random noise issuing from the loudspeakers is picked up by the microphone. It is passed through the amplifier, amplified, and fed to the loudspeakers which reproduce it louder than before. Again it is picked up by the microphone and passes through the amplifier to be reproduced louder still from the loudspeakers. The microphone picks it up once more and the cycles continue with the volume increasing each time.

The result is the familiar howl. But why not repetitions of the original sound? That merely acted as a trigger to initiate oscillation in an unstable system. Any system that has amplification and positive coupling between its input and output is unstable, and when either the amplification or the coupling reaches a critical level, it goes into oscillation. The slightest disturbance starts it off.

In this case the coupling is the air mass between the microphone diaphragm and the loudspeaker cone. If there were no coupling there would be no feedback.

So if microphone and loudspeaker were in different rooms and there were no air leak between them no feedback would occur.

Feedback rate

The feedback process happens a lot quicker than it takes to describe. Sound waves from the loudspeakers arrive back at the microphone at the speed of about 1120 ft/s (341 m/s) so the feedback cycles occur at the rate of hundreds per second. That is why the sound is a continuous howl rather than a train of individual sounds.

Once started, the feedback arrives at the microphone in a continuous stream and so is not dependent on the distance between loudspeaker and microphone to determine the howl pitch or frequency. This is governed by the dominant resonance of the system, which is usually that of either the microphone diaphragm or loudspeaker cones.

As each cycle of feedback is of greater amplitude than the previous one, it builds up rapidly, the ultimate limiting factor being the power rating of the amplifier. So, if the volume control setting is well over the critical point, feedback howl will occur at maximum amplifier power.

Thus, the operating level must always be kept well below oscillation level. But this means that the volume may be insufficient, especially with quietly spoken speakers, or those who ignore the microphone. These, like the poor, seem to be always with us! The alternative is to so design the system that feedback oscillation starts at a higher volume level; we then say that we have raised the feedback point.

Feedback reaches the microphone via two possible routes: one is direct from the loudspeakers, and the other is reflected from the walls of the auditorium; we call the latter indirect sound. The first can quite easily be prevented by using directional loudspeakers such as columns, or other types of line-source systems such as the line source ceiling array, LISCA, aimed so that their output is directed into the audience and away from the microphone.

The second route is less easy to prevent, in fact it cannot be completely eliminated. It can be minimized by arranging the loudspeakers so that no sound is radiated directly at the walls and ceilings, thereby reducing wall reflections. This is well achieved in the LISCA system.

Another effective way of reducing the pickup of indirect sound is by using super-cardioid or hyper-cardioid microphones. As direct sound comes back to the microphone at all angles, any rejection of sound from angles other than that of the desired source reduces feedback. The micropone should also be free from large peaks in its frequency response. These points will be covered in Chapters 4 and 11.

Intelligibility

Reducing feedback then, is the biggest factor in obtaining sufficient volume, but now we will take a look at the second requirement, intelligibility. This really is the most important of all; moderate or even low-volume speech that is intelligible is far better than high-volume speech that is not. So we need to understand what makes speech intelligible in order to know how to preserve it in reproduction.

Speech consists of two main parts, vowels and consonants. In general, vowels are easily recognized because they are distinctive, and the long vowels at least occupy more time than other speech sounds. They also consist mostly of the lower speech frequencies.

Consonants are less easy to identify because many are so similar, they occupy only a short time, often being little more than transients, and they are made up of the higher speech frequencies of 1–2 kHz. Yet, because there are so many more of them than vowels, they are of major importance. They distinguish between the many different words that share the same vowel sounds and so otherwise sound the same.

So, intelligibility depends to a great extent on the clear and accurate reproduction of consonants. Impaired consonants can render speech quite unintelligible. Yet it is the consonants that often suffer in a poor public address system, as we shall see. Added to this is the fact that many speakers fail to enunciate consonants adequately, especially at the ends of their words. The glottal stop, common in some parts of the country, in which the 't' sound is omitted completely, is a case in point.

The classic example of misunderstanding due to poor intelligibility is the story of a message said to have been relayed by word of mouth during a battle in the First World War. It started as: 'we are going to advance, send reinforcements'. By the time it got back to HQ it had become: 'we are going to a dance send three and fourpence'. The vowel sounds were practically unchanged, it was the consonants that suffered, so causing such a drastic change of meaning — and bewilderment to the top brass!

The ear has, among its many remarkable design features, the ability to fill in missing sounds, even complete syllables, in a familiar context so that it hardly notices that they are missing. For example, if a chairman at a meeting said: 'Good eve-ing ladies and gene--men, it gives me grea- pleasu- to int-duce our speak- for this eve-ing'. There would be little doubt that almost everyone would understand what he said. But if he went on to announce a speaker with an unusual name talking about an unfamiliar topic, missed syllables would probably mean misunderstanding.

This is a problem when trying to assess the intelligibility of a public address system. When hearing a test passage read over the system, how can it be determined what was actually heard and how much was aurally filled in?

The solution employed in standard intelligibility tests is to use monosyllabic nonsense words. A number of volunteers sit in various parts of the auditorium and write down what they think they heard, while a speaker reads out a list of nonsense words. The degree of accuracy gives the PSA (percentage articulation index) of the system. A 100% is unheard of and never attained; 95% is considered

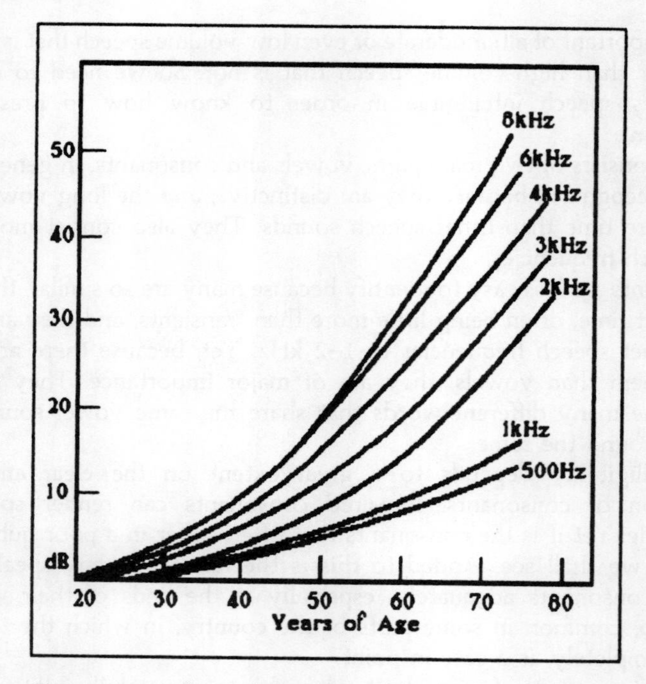

Figure 1 Average hearing loss with age. High frequencies suffer most, thus affecting speech intelligibility.

the maximum and not often achieved. Around 80% will result in reasonably good audience comprehension, but at 75%, concentration will be necessary. Below 65% indicates poor intelligibility. A realistic aim is thus to get the PSA somewhere between 80% and 95%, above 90% if possible.

Another factor affecting intelligibility is hearing impairment due to the age of the listener, known as presbycusis. Hearing declines with age, but more so at the higher frequencies. The aural response of a 50-year-old to frequencies of 4 kHz is some 10 dB down on what it was at 20. At 60 this loss becomes about 24 dB, while for 2 kHz it is 12 dB and for 1 kHz around 6 dB. The low frequencies are relatively unimpaired (Figure 1).

Deterioration also occurs if the subject has been exposed to long periods of excessive noise, at work or elsewhere, but irrespective of the type of noise, the impairment starts at 4 kHz, and spreads lower.

So the higher speech frequencies, which are the ones most needed to reproduce consonants so necessary to achieve good speech comprehension, are the ones most affected both by advancing age and hearing damage.

From this it is evident that although public address systems that distort the high-frequency band or emphasize the low will have generally poor intelligibility, some in the audience will be more affected than others. Many, especially the young, may be unaware of any deficiency, but older listeners may experience considerable difficulty in comprehension.

It can be seen from all this that it is not easy to judge how good a system is by just giving a short listening test. It may seem quite good to an under 40-year-old with a speaker who articulates well, but how will it sound with a poor speaker to an older person? One way to assess it is to conduct a PSA with persons of various ages, but rather than finding deficiencies after the system is installed, it is better to design the system along good acoustic principles from the start.

Naturalness

Natural sounding public address is a goal to be achieved if at all possible. With large installations it is less easy to attain than with smaller ones which strictly speaking are *sound reinforcement systems*, but much can be done in both cases to make the sound natural. Ideally it should seem as if the sole source of the sound were the speaker, except that the volume is sufficient for everyone to hear.

In small or medium-sized halls the first couple of rows need only moderate sound reinforcement, because the speaker is usually only a few feet away, and his natural voice can be heard directly. Such reinforcement should appear to come from the direction of the speaker, so the source should ideally have a centre-front location. Further back, there is little direct sound from the speaker and all comes from loudspeakers, but it still should sound as though it were coming from the platform. A frontal source is therefore essential, and a central location is the same plane desirable.

This is not easy to achieve with conventional systems. With column speakers, a frontal source location in the same plane is indeed obtained, but for the majority of the audience, the source is at the side, where the columns are usually installed. For the ceiling matrix system matters are much worse as the sound comes from overhead and for many places in the audience, from behind. The LISCA system, described later, does give front-centre sound everywhere, and the desired reduced volume for the first rows.

In addition to natural source location, the sound should have a natural tone, being free from boominess, harshness or other tonal defects. Apart from possibly impairing intelligibility, such deficiencies make the audience constantly aware of the sound system rather than the speaker.

It may be thought that as long as the sound can be heard and understood, that is all that matters; the location and tone are unimportant. While they are secondary, they are nonetheless important. Deviation from natural sound can cause subconscious mental fatigue, and reduce the attention span of the audience. So, the quality of naturalness should never be underestimated.

Reliability

This is an essential requirement as a system failure could be catastrophic at a large gathering attended by thousands. It could even be life-threatening if the system were being used to relay warning messages in the event of an emergency.

Reliability is not something that just happens, or can be hoped for without planning for it. It has to be designed into the system and all due precautions taken when installing it. Furthermore, there should be stand-by equipment, test procedures, and contingency plans evolved on a 'what if?' basis. These points will be looked at in detail in Chapter 10.

There then we have the basic requirements; we can now go on to see how they can be achieved in the design of the system. Firstly, though, we need to have a good grounding in the basic principles of acoustics and how sound behaves in a typical auditorium as well as at open-air sites. This will be covered in the next chapter.

2 Acoustics and the nature of sound

It has often been observed that some public-address installers and operators, though having good knowledge and experience of electronics and sound equipment, have scant understanding of the principles of acoustics. This frequently shows up in installations that have an ineffective loudspeaker system or unsuitable microphones.

The way sound pressure waves behave and how they are affected by the auditorium has an obvious and considerable effect on feedback, intelligibility and sound coverage. So, for successful public-address work a thorough working knowledge of *both* the electrical side and acoustic principles is essential.

The study of acoustics in all its aspects is a big subject which cannot be covered in a single chapter. So, we will consider here those features that have particular relevance to public-address installations. For the sake of newcomers to the subject, we will start with basics which we hope professional readers will bear with, and continue from there.

Sound waves

The term *wave*, although correct, can be misleading. We usually think of waves as those seen at the seaside, vertical displacements of water consisting of ridges separated by troughs.

Sound waves, although behaving in a similar manner to sea waves or ripples on water in the way they are propagated, diffracted and reflected, are not vertical variations but consist of backwards-and-forwards motions of the air particles. These produce successive regions of compression and expansion or pressure differences, which spread outward from the source.

Air is a springy material, as anyone who has tried to operate a blocked air pump will have discovered, so the progress of sound waves can be illustrated by imagining a long coiled spring supported at its ends. A series of longitudinal impulses applied at one end travel along it as a train of compressions and expansions between the individual coils as shown in Figure 2.

Just as the coils of the spring do not travel from one end to the other but move backwards and forwards, so the air particles themselves do not move outward from the source, but each imparts oscillatory motion to the next.

Wavelength and frequency

Wavelength is obviously the distance from the crest of one wave to that of the next — or for that matter any part of a wave to the corresponding part of its successor. When a series of waves is travelling through a medium at a fixed speed the number of waves that pass a given point in a given time can be

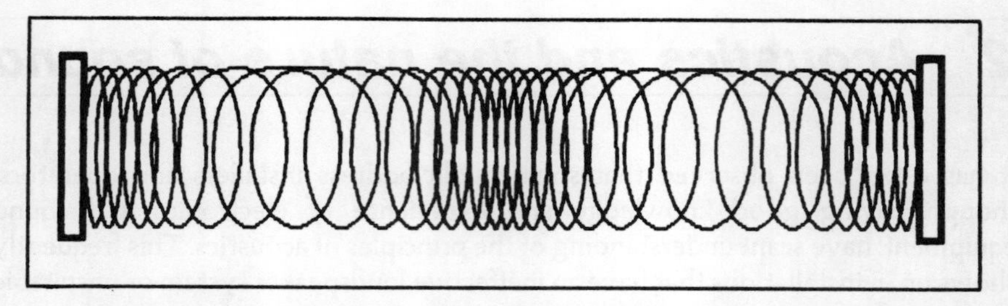

Figure 2 A sound wave travels through air like a compression wave along a spring producing travelling regions of high and low compresson.

counted. This we term the *frequency*. If the wavelength is short, there are many waves in a given area, and a large number pass the given point in the specified time. If, however, the wavelength is long, say twice as long as before, there will be fewer waves, in this case half the number.

Frequency was once described by the term cycles-per-second which may still be encountered, the unit being written as c/s. The common multiple is the kilocycle-per-second being a thousand c/s. The cycle-per-second has now given way to the *hertz* (Hz), the multiple being the *kilohertz* (kH).

Wavelength has no unit other than physical length, usually expressed metrically. So we say that a certain sound frequency has a wavelength of so many metres, or if a high frequency, so many centimetres. It is often depicted by the Greek letter lambda (λ).

It should be noted that the relation between a particular frequency and its wavelength is not absolute but is governed by the speed of the wave through the medium. For sound this is 1120 ft/s (341 m/s) for a temperature of 60°F (15.5°C); it increases by 1.1 ft per second for a temperature rise of each degree F(0.61 m per second for each degree C). It is also affected by barometric pressure, but the effect is much less.

The concept of wavelength and frequency and their relationship is fundamental to all matters acoustic, so should be thoroughly understood.

Intensity and pressure

Sound intensity is the amount of acoustic power passing through a given area, usually 1 square metre. As with electrical power the unit is the watt. It is usually expressed as a ratio of the faintest sound that can be heard — the hearing threshold; at 1 kHz, this is 10^{-12} W, which deflects the human eardrum by less than the diameter of a single atom.

Sound intensity is used as a measure of total energy generated by a source such as a piece of machinery which it is required to silence, but it is not so often used for general sound measurement. For this, the pressure makes a more convenient factor as it relates to the effect sound has upon our eardrums. Many microphones (though not all) are likewise actuated by sound pressure waves.

The sound pressure level is usually denoted by the letters SPL, and if we compare it with electrical terms, it corresponds with the voltage in a circuit. As voltage is the square root of wattage in all circuits having the same impedance, so SPL is the square root of intensity.

Various units have been used to quantify pressure. These are the bar, which is the atmospheric pressure at sea level; the dyne/cm²; the newton/m²; the pascal; and the lbf/in² (psi). The relation between these is:

$$10 \text{ dyne/cm}^2 = 10\ \mu\text{bar} = 1 \text{ newton/m}^2 = 1 \text{ pascal} = 0.000145 \text{ lbf/in}^2$$

The pascal is now the preferred unit.

The decibel

The decibel is often incorrectly considered to be the unit of sound level. It is not, but express the ratio between two different signal or sound levels. When applied to a sound pressure level it denotes the ratio between that level and the threshold of hearing which is 20 μpascals ($20 \log_{10}$). When applied to a sound power or intensity level it is the ratio between that power and 10^{-12} W ($10 \log_{10}$).

It is sometimes asked: why use a ratio instead of some absolute unit for measurement? The answer is the way the human ear reacts to different sound pressures. It judges sound levels by their ratio to each other and its perception is not linear but logarithmic.

We have a sort of automatic volume control in our ears whereby the pivotal positions of the small bones in the middle ear, which convey the vibrations from the ear drum to the cochlea, vary according to the amplitude of the sound. They are in the position for the maximum transfer of energy with very quiet sounds, but in the position for minimum transfer with loud ones. This remarkable design feature gives us a huge dynamic range, allowing us to hear the rustling of a leaf and also the roar of a thunder clap.

However it complicates sound measurements if they are to relate to what we actually hear. Hence the use of the decibel, which is a logarithmic ratio. When it is applied to sound pressure level it strictly should have the suffix SPL to distinguish it for the sound power level, but the letters are customarily dropped. The threshold, having a ratio to itself of unity is denoted by 0 dB since log $1 = 0$, while the loudest sound that can be heard without hearing damage is 120 dB.

Twice a particular level is 6 dB; four times, 12 dB, eight times 18 dB and so on on. A change of 1 dB is the smallest level that can be detected.

Table 1 Ratios and dB values

dB	ratio	dB	ratio	dB	ratio	dB	ratio
0	1.0	2.5	1.334	10.0	3.162	50	316
0.1	1.012	3.0	1.413	11.0	3.55	60	1000
0.2	1.023	3.5	1.496	12.0	3.98	70	3162
0.3	1.035	4.0	1.585	13.0	4.47	80	10^4
0.4	1.047	4.5	1.679	14.0	5.01	90	3.16×10^4
0.5	1.059	5.0	1.778	15.0	5.62	100	10^5
0.6	1.072	6.0	1.995	16.0	6.31	110	3.16×10^5
0.8	1.096	7.0	2.239	18.0	7.94	120	10^6
1.5	1.189	8.0	2.512	20.0	10.0		
2.0	1.259	9.0	2.818	40.0	100.0		

Weighting curves

The ear is not equally responsive to all frequencies nor at all sound levels. This is shown in Figure 3 which are equal loudness contours. They show the amount of sound pressure required to produce sensations of equal loudness at various frequencies and volume levels. They are therefore the inverse of frequency response curves.

It is thus considered necessary to relate the frequency response to the level of measured sound in order to reflect its aural effect. This is done by using weighting curves as shown in Figure 4. The one used for public address work and in fact most applications is the *A* curve and measurements made according to it are thus designated as dBA, although the *A* is often dropped.

Phase

Another important factor which affects the behaviour is its phase. As we have seen the sound wave consists of regions of alternate high and low pressure. We have a region of compression followed by a rarefied one. This can be visualized by the action of a loudspeaker cone. When it moves forward the air in front is compressed, then when it moves backward, the air is rarefied. The two waves are thereby said to be out of phase. If two out-of-phase waves mix, the troughs and peaks cancel each other and there is zero sound pressure.

This happens at the rim of an unmounted loudspeaker, all wavelengths longer than the radius of the cone suffer cancellation, but those that are shorter are unaffected. This is responsible for the tinny effect of an unmounted loudspeaker.

When two sound waves mix, they may be in phase and so reinforce each other, or they may be out of phase and thereby cancel. It is also possible that

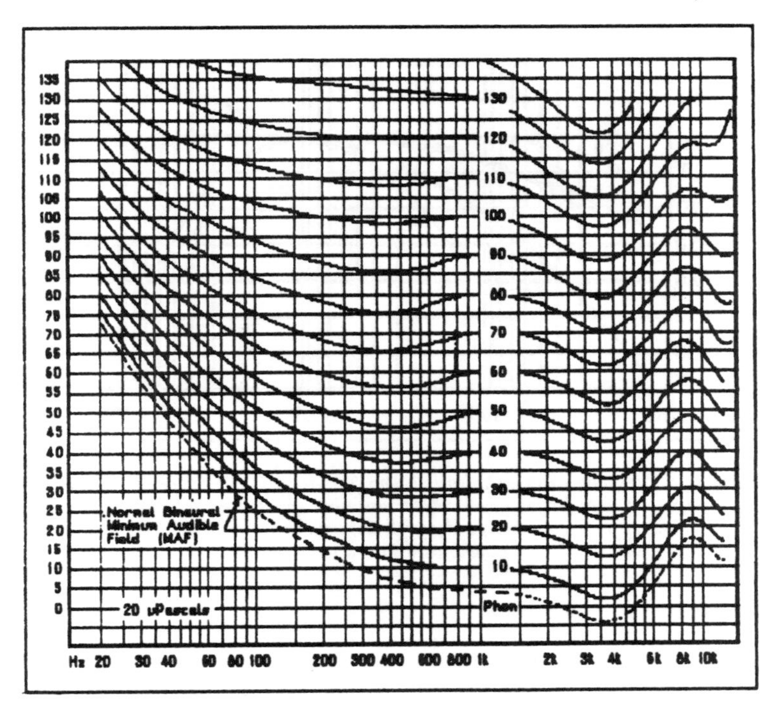

Figure 3 Equal loudness contours showing the SPL required at various frequencies and levels to produce sensations of equal loudness. They are thus the inverse of aural frequency response.

they may be only partly out of phase, with the compression region of one coming part way between the compression and rarefied regions of the other. For this reason phase differences are represented by a circle and the various sound waves as radii of the circle at different angles. When two are exactly out of phase, they are represented by two radii opposite each other to form a straight line bisecting the circle. They are thus said to be 180° out of phase.

Other phase differences are also expressed as degrees of a circle. A quarter cycle displacement is thus 90°, and the signals are said to be in quadrature. Now if two in-phase waves of equal amplitude reinforce and so produce double the pressure, while two waves that are 180° out of phase cancel to give zero pressure, it follows that some in-between phase angle will produce a pressure that is in between those two extremes.

The resulting pressure and its phase (termed the *resultant*) can be determined by a little geometry. All we have to do is to draw a line representing one value from the end of a line representing the other at the appropriate phase angle, both lines being to scale. A third line drawn from the two free ends to form a triangle represents the resultant, its length giving the amplitude to the same scale (Figure 5). The lines are called *vectors*.

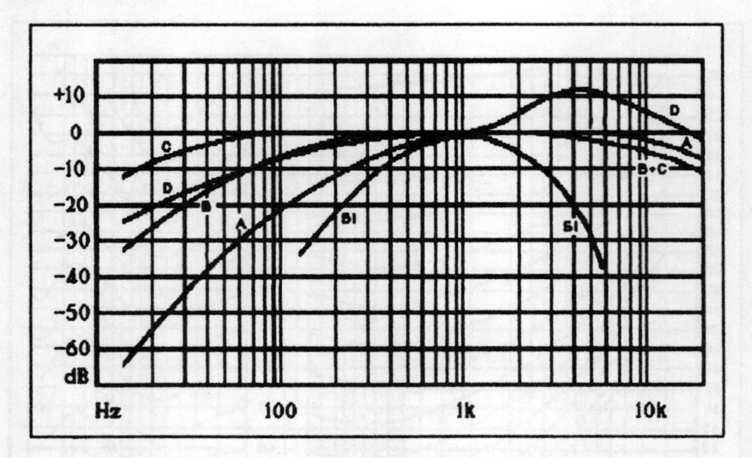

Figure 4 Weighing curves: A, levels below 55 phons; B, 55–85 phons; C, above 85 phons; D, for aircraft noise; SI, for assessing speech frequency interference. B and C curves have not corresponded well with subjective results so the A curve is now used for all levels.

Interference

In a hall or auditorium where loudspeakers are operating there are numerous reflected waves as well as direct ones from the loudspeakers. Many of these combine at various points to produce reinforcement or cancellation. This is termed interference. When two waves converge after travelling along different paths of different lengths from the same source, there are phase differences. But these are not constant, they vary with the wavelength of the sound.

For example, a difference in length of 6.75 (17.15 cm) inches between two paths is half the wavelength of 1 kHz, so cancellation occurs. When this distance

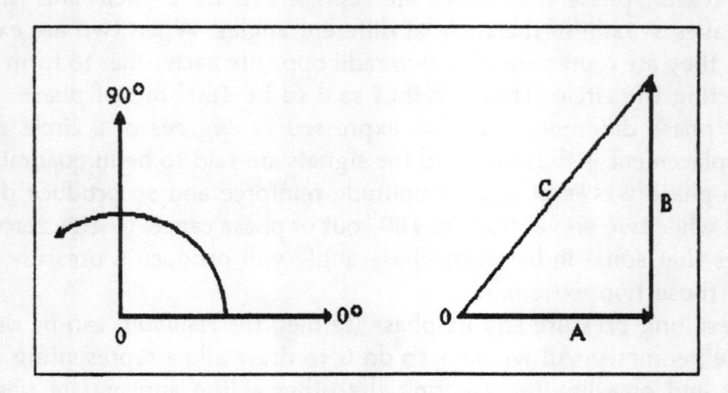

Figure 5 Vectors, signal amplitudes are indicated by line lengths and phase by angles. Rotation is 360° for a complete cycle. If two signals A and B are drawn one from the end of the other, the phase and amplitude of the resultant is shown by the joining line C.

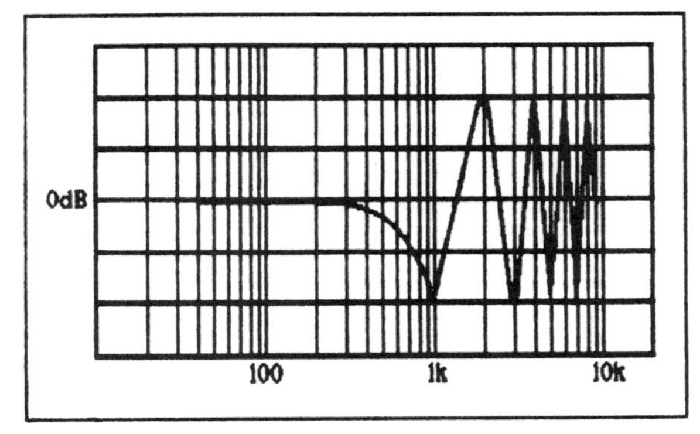

Figure 6 Comb filter effect. With a sound path difference of 6.75 ins (17.15 cm) then response falls to total cancellation at 1 kHz, then reinforcement at 2 kHz and so on.

equals $1\frac{1}{2}$ times the wavelength which is at 3 kHz, cancellation also takes place, as it does at $2\frac{1}{2}$, $3\frac{1}{2}$, $4\frac{1}{2}$ times (5 kHz, 7 kHz, 9 kHz) and so on.

At length differences of 1, 2, 3, 4 times the wavelength (2 kHz, 4 kHz, 6 kHz, 8 kHz) there is reinforcement. The frequency response thus oscillates violently between successive peaks and troughs after a gradual drop from the bass to the first cancellation at half-wavelength (Figure 6). The peaks resemble the teeth of a comb, hence the description: *comb filter* effect.

As the affected frequencies depend on the path length differences between two reflections or between two sources, they vary at different positions in the auditorium. Comb filter effect is more complex when more than two waves arrive at different intervals, such as at an off-centre point within a quadrant of ceiling loudspeakers.

The effect produces frequency distortion at high frequencies which are the ones that convey speech consonants. They are thereby rendered indistinct and ambiguous, so intelligibility suffers. Strong reflections or multiple sources are therefore to be avoided. This is why the British Standard BS 6259 for sound distribution systems stipulates that sound should come from one virtual source if at all possible. It is also why the vogue for rows of ceiling loudspeakers in matrix form is unsatisfactory.

Another area where comb filter effect could be encountered is near a microphone. If there is a hard reflective surface nearby such as a table top, sound can reach the microphone direct from the source, and also reflected from the surface. The effect is even worse than with multiple loudspeakers, because the whole system being fed from the microphone is affected, whereas only parts of the auditorium are affected by loudspeaker mutual interference.

Monopole source

A point source that radiates equally in all directions is termed a monopole. The sound wave travels outward in the form of expanding concentric spheres. As

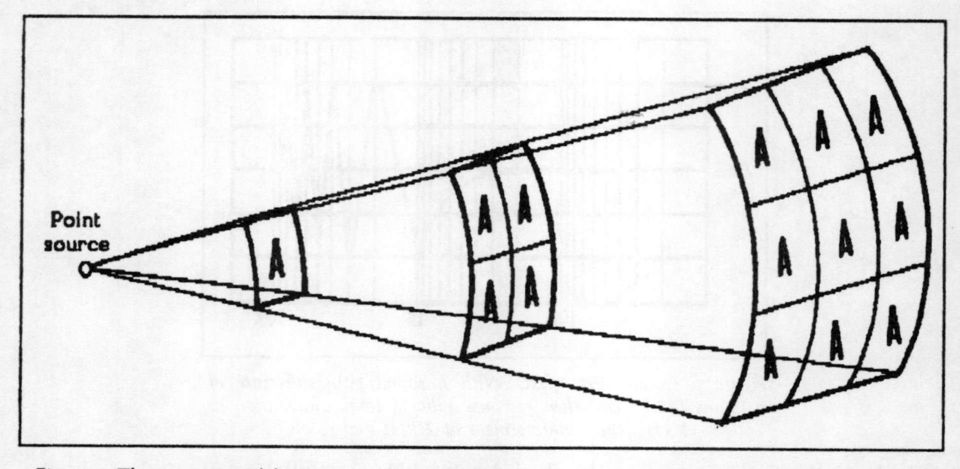

Figure 7 The area covered by an expanding sound wave is the square of the distance from the source, hence intensity follows an inverse square law.

the sphere expands, the energy at any section decreases because it is spread over an increasing area.

Figure 7 shows a segment of a sphere, and at a distance from the point source there is an area *A*. At twice the distance, the area has increased four times, while at three times the distance it has multiplied to nine times, so following a square law. The energy or power therefore decreases by the same amount and obeys an inverse square law, becoming a quarter at twice the distance, a ninth at three times, and sixteenth at four times and so on.

This is true of the sound *power* or energy, but as we have seen, the sound *pressure*, which is the factor normally used in measurements, is the square root of the power. Hence the pressure is inversely proportional to distance from the source.

Drop in sound pressure as a function of distance is thus due almost entirely to expansion of the wave. Apart from this, sound can travel considerable distances with little loss. There is a slight attenuation due to air friction loss which is greater at high frequencies because there are more air particle oscillations, hence friction, at high frequencies than low. At 1 kHz there is a 2.5 dB drop over 1000 ft (305 m); at 2 kHz the drop is 5 dB over 500 ft (152 m). At 3 kHz there is a 10 dB loss over 500 ft (152 m), while at 10 kHz the loss is 10 dB over 100 ft (30.5 m). (Figure 8).

In most auditoriums it can be seen that the loss at speech frequencies is slight and is due mostly to sound wave expansion from the source. A loudspeaker in a sealed box is a monopole at all wavelengths greater than the largest dimension of the box.

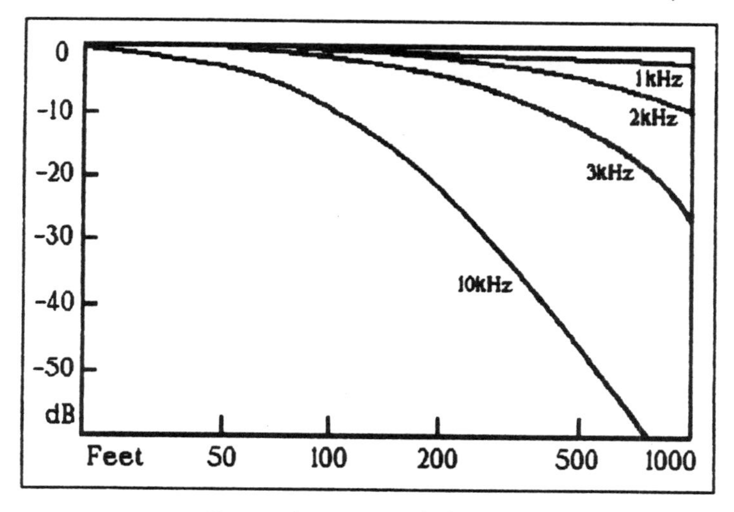

Figure 8 Attenuation with distance.

The dipole

The dipole or doublet has a distribution pattern in the shape of a figure 8, with two circular lobes 180° out of phase, one at the front and the other at the back. The sound pressure level (SPL) at any point depends not only on the distance from the source, but also the angle from the main axis. This is an important factor that is involved in public address loudspeaker system design as most loudspeakers with non-sealed backs are doublets.

The polar diagram is shown in Figure 9, and the sound pressure level at any point off-axis is the cosine of the angle multiplied by the pressure on-axis at the same distance.

Table 2 Sound pressure level values for off-axis angles

0°	1.000	25°	0.906	50°	0.643	75°	0.259
5°	0.996	30°	0.866	55°	0.573	80°	0.174
10°	0.985	35°	0.819	60°	0.5	85°	0.087
15°	0.966	40°	0.766	65°	0.423	90°	0
20°	0.939	45°	0.707	70°	0.342		

It is worth noting that the SPL drops only slightly to start when moving off-axis, but decreases more rapidly thereafter. It is useful to remember that the half SPL point (−6 dB) is at 60° off-axis. As with a monopole, the attenuation with distance is 6 dB for a doubling of distance from the source.

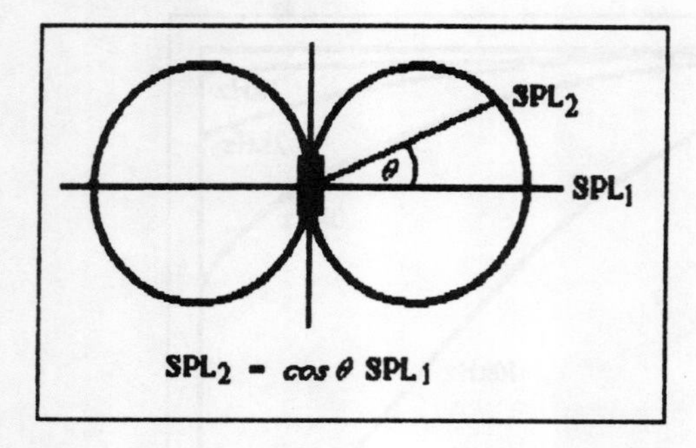

Figure 9 Polar diagram of a doublet.

Line source

A line source radiating sound throughout 360° does so in the form of expanding concentric cylinders rather than spheres. The sound energy is thereby concentrated within the horizontal plane with very little radiated vertically (Figure 10). Attenuation with distance is thus less than for a monopole or dipole. It is 3 dB for each doubling of distance, and 6 dB for quadrupling. It thus carries twice as far as the other sources.

The principle is used in the column loudspeaker which has a vertical array of drivers, but it is not a true omnidirectional line source as the propagation is restricted at the back and sides. The polar response seen from above is as shown in Figure 11. The lack of vertical radiation produces a flat topped and bottomed beam of sound similar to a dipped headlamp beam.

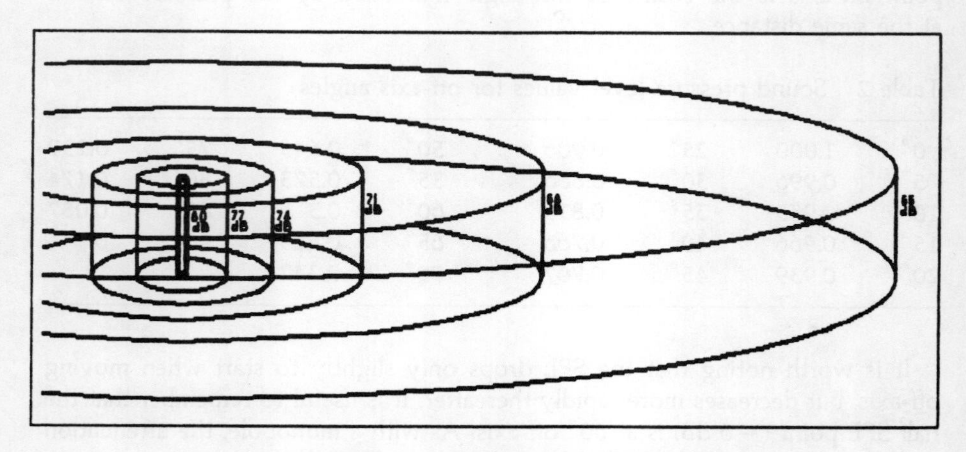

Figure 10 Line source, attenuation is 3 dB for doubling of distance.

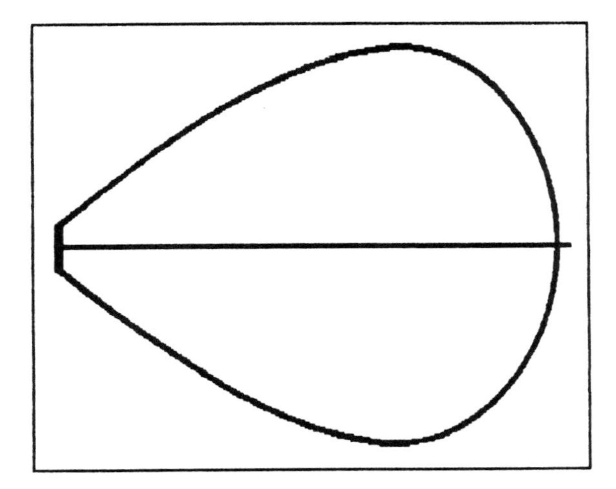

Figure 11 Polar diagram of a column speaker.

Restricted propagation

If a sound wave is confined to a tube or restricted channel, it is prevented from expanding and so does not suffer the normal decrease in pressure. Losses are due only to air turbulence and friction, so sound energy can be transferred over quite lengthy distances with little attenuation. Examples are: acoustic speaking tubes, noisy air heating ducting, and sound trapped in narrow air channels due to temperature inversions.

Plane wave fronts

Most sources radiate sound as expanding concentric spheres or cylinders or segments thereof. Close to the source, the wave has a curved front, but as it progresses, the sphere becomes larger and the arc of any given segment less curved. Eventually at some distance from the source, the wave front becomes virtually flat or plane. The effects of spherical and plane wavefronts can be quite different. This is especially so in their effect on hearing; the source location of spherical waves can be more readily identified than those of plane waves.

When a sound wave meets an obstacle, it can be reflected, refracted, diffracted, absorbed, or transmitted through it. Usually it is some of each, but the dominant effect depends on the type of material and its size. We will consider each of these five effects in turn.

Reflection

Reflection occurs when the surface is hard and smooth. Ceramic tile on concrete is one of the best reflectors, and bare concrete and stone follow closely. Brick

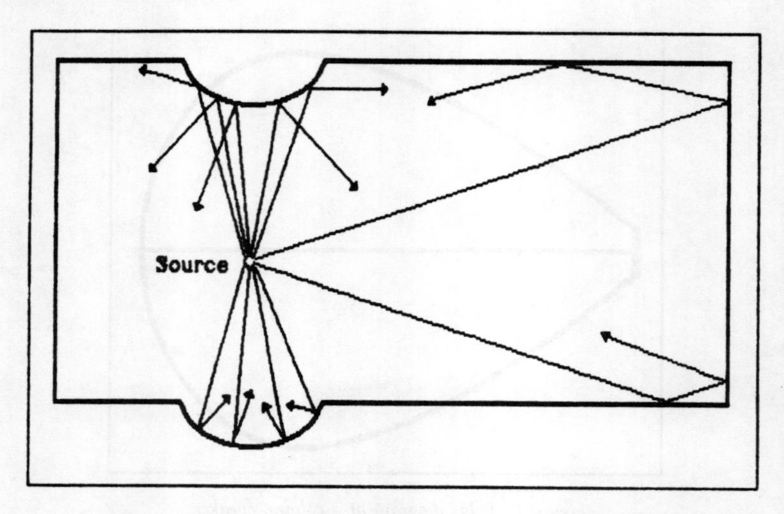

Figure 12 How sound is reflected from different shaped surfaces.

and plaster on brick are only slightly less so. Wood is also reflective, especially if polished; the thicker the wood, the more reflective it is, but the thickest wood is not so reflective as brick.

When a sound wave meets a reflective surface, it behaves in a similar manner to light. The angle of reflection equals the angle of incidence. So, if a sound wave arrives at 90° to the surface it is reflected straight back to the source. If it comes in at an acute angle, it is reflected off at the same angle. A sound wave directed into a right-angled corner always comes out parallel to the inward path irrespective of the angle of incidence.

When impinging on a convex surface it is scattered outward on divergent paths, but when encountering a concave surface it converges. These various angles are illustrated in Figure 12.

Another effect of reflection is that a ghost image is created behind the reflecting surface at a depth equal to the distance that the source is in front of it (Figure 13). Comb filter effects are also produced.

Diffraction

Each point on a pressure wave front tries to expand in all directions as if it were a new source. However, it cannot expand sideways because of the similar pressure of adjacent points. It can thus only go forward to areas of lower pressure.

When a sound wave encounters and passes the edge of an obstacle, there is no side pressure from adjacent wave points and so the wave is able to expand sideways. It thus flows around the object, and the effect is termed diffraction (Figure 14a), but it does this only if the wavelength is large compared to the

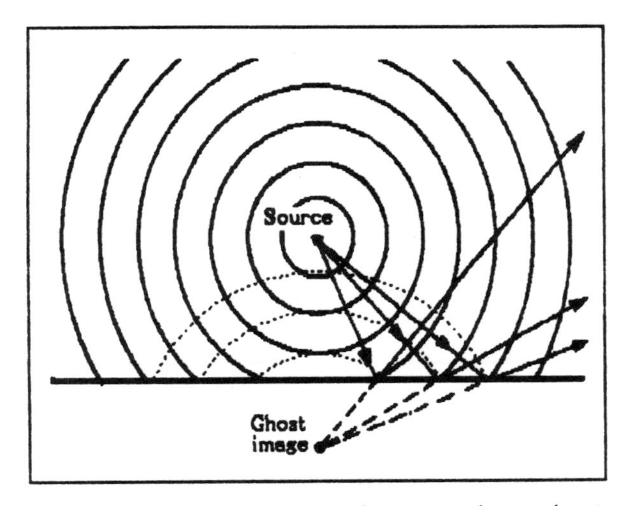

Figure 13 Reflections from a surface near the source produces a ghost image.

largest dimension of the object. If the object is larger, then it casts an acoustic shadow (Figure 14b). Diffraction does not commence suddenly when the wavelength reaches a critical point, it gradually increases with wavelength.

At a frequency of 246/*d*, in which *d* is the dimension in feet (75/*d* with *d* in metres) of the shortest side, the SPL is −3 dB behind the object. At a frequency of 985/*d* (or 300/*d* if *d* is in metres), the SPL is −10 dB.

It follows from this that objects reflect short wavelengths that are not diffracted, but not long wavelengths which flow around them. Thus beams and other

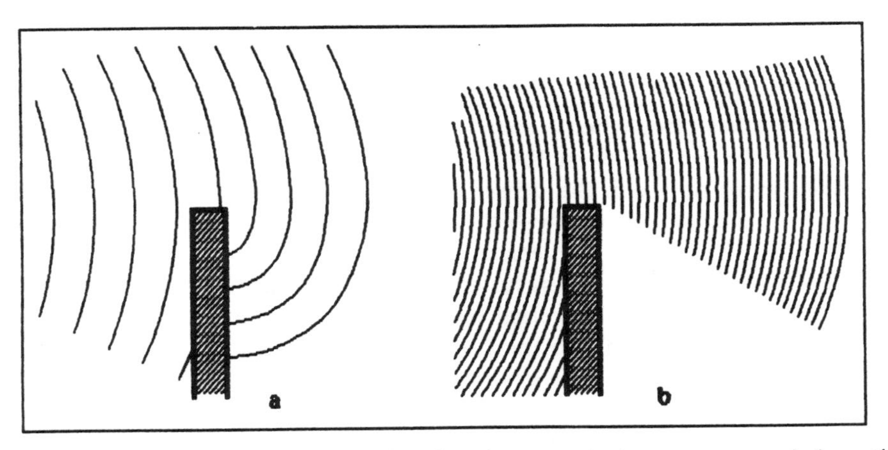

Figure 14 (a) Long wavelengths are diffracted behind an obstruction; (b) obstruction causes a shadow with short wavelengths.

structural projections can reflect high frequencies and thereby contribute to feedback. This fact must also be considered when designing deflectors or baffles, they must be large enough to be effective at the lowest required frequency, otherwise those frequencies will just bypass them.

Diffraction also occurs when a sound encounters a hole, wavelengths that are long relative to the size of the hole pass through and diffract around it in a hemisphere. Even when the oncoming sound is a plane wave, the wavefront radiating from the hole is spherical as if the hole was itself a new source. When the wavelength is short compared to the hole size, it passes through as a beam and does not diffract (Figure 15).

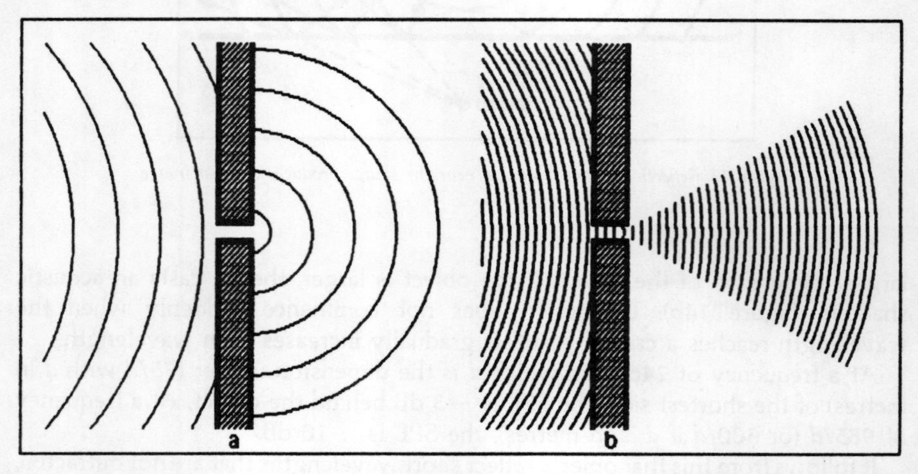

Figure 15 (a) Wavelengths larger than a hole are diffracted around it forming spherical wavefronts; (b) at short wavelengths diffraction does not occur so producing a sound beam.

Refraction

Refraction takes place when a sound wave passes at an angle from a medium of one density into one of another. This is due to the different sound velocities, it travels more slowly in the denser medium and so bends into it. It is rather like the nearside wheels of a vehicle that encounters pools of water on a wet road. They tend to slew the vehicle to the nearside.

In the case of a public-address system, refraction occurs when the sound waves pass through air layers of different temperatures. Cold air is denser than warm so the sound is refracted into the colder region. Usually the air near the floor is cold while that near the ceiling is warm. Sound waves propagated from vertical columns thus tend to be bent downward into the audience and away from the ceiling and upper walls, a helpful characteristic.

With open-air public address, the sound bends upward because temperature normally gets cooler from ground level upwards; there is what is known as a

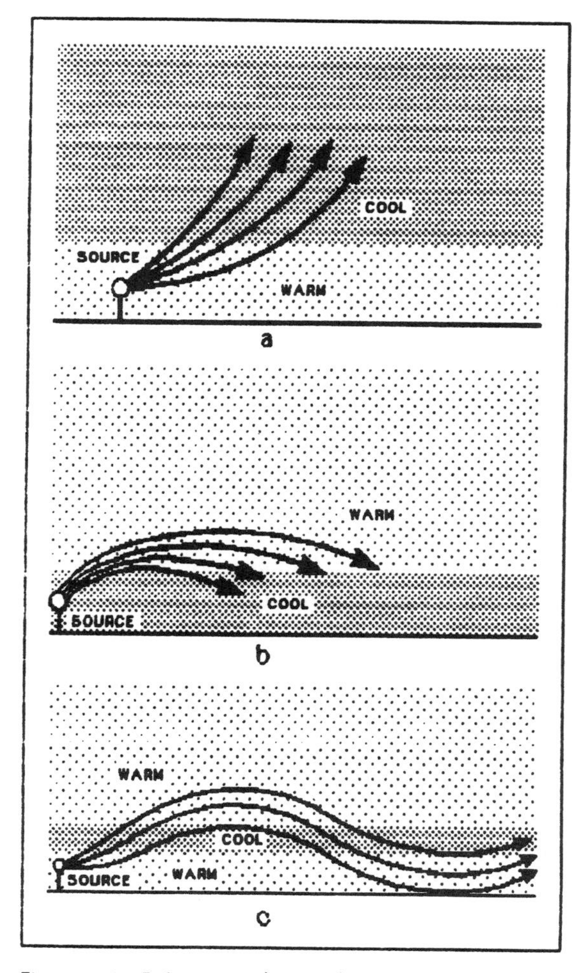

Figure 16 (a) Refraction with normal temperature gradient; (b) inverted temperature gradient; (c) a double gradient, sound is trapped between layers and can travel considerable distances.

temperature gradient (Figure 16a). Thus sound does not travel very far over level ground.

It is interesting to note that ancient amphitheatres were built in ascending tiers and in semicircular form. They thus followed the natural path of the sound waves both horizontally and vertically. Furthermore, being built of stone, they were non-absorbent and so gave excellent sound distribution over the whole audience.

Occasionally, there is a *temperature inversion* when the air is cooler at ground level than higher. Sound is then bent downward and travels along the surface of the ground instead of being radiated upward. It thus can carry for a considerable distance (Figure 16b). This condition is often met over water which cools the adjacent air layer. A further layer of air at normal temperature, which is warmer, lies above this. So sound can carry for surprising distances over lakes and rivers, and if calm, over the sea. An example of this is in the Biblical accounts of Jesus Christ addressing large crowds from a boat moored just off the shore. He was evidently well aware of the effect on sound of temperature inversion.

On rare occasions, a *double gradient* occurs when a warm air layer lies over the normal cool one, which in turn is above the warm ground air. This gives a sandwich effect of a cool layer between two warm ones. Sound is thereby trapped and is refracted from one to the other (Figure 16c). This can result in the sound being heard more clearly at a distance than nearby.

Effect of wind

Wind velocity increases with height above the ground, being theoretically zero at ground level, although this is rarely so in practice. When propagated against the wind, the sound path is bent upward. This is because the oncoming rush of air well above ground level is at a higher pressure, hence is denser, than the calmer layers below it. As with refraction, the sound is slowed in a denser medium and so is bent upward towards it. When propagated in the same direction as the wind, the receding upper layer is less dense than the slower moving layer beneath, so the sound path is bent downward and kept at ground level.

The effect is that sound is heard at a greater distance with the wind than against it. This is common knowledge, but the reason for it is less well known. The sound is not 'carried' by the wind as is popularly supposed but bent earthward toward downwind listeners. All these factors could be significant with outdoor public address systems, so the topography and prevailing wind direction should be considered when planning such an installation.

Absorbents

Absorption results when a sound wave encounters materials that are soft and thick. Curtains, upholstery, furnishing, carpets, and clothing are all common absorbents. Acoustic panels and tiles are highly absorbent and are used where sound deadening is required. The energy is converted into heat, but it is of too low an order to be measured. Some sound is reflected, and some transmitted through the material.

Contrary to what may be expected, absorbents are not always good sound insulators. Their soft porous nature, though non-reflective, permits sound to travel through them unless very thick. It is the hard reflective surfaces that are usually the best insulators of airborne sound as they reflect most of it away. Things are different for sound sources in physical contact with the medium, in this case hard objects conduct sound ready through them, while the soft ones dampen it.

Materials are rated according to their absorption coefficients from 1.0 which is 100% absorption, equivalent to an open window, to 0 which denotes 100% reflection. Absorption coefficients vary with frequency; most have greater absorption at high frequencies than low, but there is one exception worth noting; this is plywood plain or veneered on 2 inch (50 mm) battens, which has good low-frequency absorption, but less at high frequencies.

Table 3 Reflective and absorbent materials

Material	125 Hz	250 Hz	500 Hz	1 kHz	2 kHz	4 kHz
Concrete	0.01	0.01	0.02	0.02	0.02	0.03
Brick	0.024	0.025	0.03	0.04	0.05	0.07
Door, wooden *	0.3		0.15		0.05	
Floor, wood on joists	0.15	0.2	0.1	0.1	0.1	
Floor, wood block	0.05		0.05		0.1	
Glass	0.03	0.03	0.03	0.03	0.02	0.02
Plaster on brick	0.024	0.027	0.03	0.037	0.039	0.034
Plaster on lathe	0.3		0.1		0.04	
Plasterboard	0.3		0.1		0.04	
Plywood $\frac{3}{8}$ in. (95mm)	0.11		0.12		0.1	
Plywood $\frac{3}{16}$ in. (4.8 mm) on 2 in. (50 mm) battens	0.35	0.25	0.2	0.15	0.05	0.05
Wood, $\frac{3}{4}$ in. (19 mm) solid	0.1	0.11	0.1	0.06	0.06	0.11
Acoustic panels	0.15	0.3	0.75	0.85	0.75	0.4
Acoustic tiles $\frac{3}{4}$ in. (19 mm) thick $\frac{3}{16}$ in. (4.8 mm) holes; $\frac{1}{2}$ in. (13 mm) centres	0.1	0.35	0.7	0.75	0.65	0.5
Carpet, thin	0.05	0.1	0.2	0.25	0.3	0.35
Carpet thick with underlay	0.15		0.35		0.5	
Chair, padded dining *	1.0		2.5		3.0	
Chair, upholstered *	2.5	3.0	3.0	3.0	3.0	4.0
Curtains, light	0.05	0.12	0.15	0.27	0.37	0.5
Curtains, heavy with folds	0.2		0.5		0.8	
Fibreglass 1 in. (25 mm)	0.07	0.23	0.42	0.77	0.73	0.7
Fibreglass 2 in. (50 mm)	0.19	0.51	0.79	0.92	0.82	0.78
Fibreglass 4 in. (100 mm)	0.38	0.89	0.96	0.98	0.81	0.87
Person seated *	0.16	0.4	0.46	0.46	0.5	0.46

Table 3 lists firstly highly reflective materials, then those highly absorbent. Those marked * are for one item, all others are per square foot for calculating total absorption. Otherwise the figures can be just used for comparison of different materials.

Special absorbers are often used in recording studios consisting of panels supported by their edges away from a wall. These vibrate at a certain resonant frequency, thereby absorbing energy at that frequency. However, they re-radiate sound so reducing effectiveness as an absorber to a coefficient of less than 0.5. Introducing damping material between the panel and the wall increases absorption and broadens its frequency range.

If the panel is perforated with small holes, resonators are formed consisting of each hole and the air gap behind it. The deeper the gap, the lower the effective resonant frequency. Varying the size and spacing of the holes spreads the frequency range. Further broadening is achieved by filling the gap with fibreglass.

The frequency can be lowered by placing a membrane such as roofing felt between the panel and the filling. A double layer reduces it further. This can be more convenient and is less space consuming than increasing the gap to lower the frequency.

Transmission

Most materials transmit or conduct sound through them as well as reflect and absorb. The high reflectors transmit less because most of the sound is in fact reflected away. However, if the source is in physical contact, reflective materials, being usually hard, conduct sound readily. Some idea of their contact conductivity can be seen from Table 4.

Table 4 Attenuation of sound through hard materials

Material:	Steel	Brick	Concrete	Wood
Attenuation dB per 100 ft (30 m)	0.3–1.0	0.4–4.0	1.0–6.0	1.5–10.0

Window glass attenuates airborne sound by around 25 dB, standard double glazing adds about 4 dB. Optimum gap between panes for sound attenuation is 4 in. (100 mm); this produces a much greater attenuation, but is not the optimum for heat insulation. Where both heat and sound insulation are important an extra pane spaced 4 in. outside a standard double glazed assembly should be used.

Reverberation

When a source ceases to propagate sound in an enclosed area, it does not stop but bounces around the area boundaries for a while until it dies away. The time it takes for a sound pressure level to diminish to one thousandth of its original level (−60 dB), is called the reverberation time. It has a major effect on the quality and intelligibility of speech. This can be demonstrated by making a recording of some speech in a bathroom, then making another recording of the same passage in a bedroom.

On comparing them the sound quality will be discerned to be totally different. That made in the bathroom will sound lively, but hollow and not always easy to follow, while the one made in the bedroom will sound dead, but immediate with every word understandable. The bathroom has a long reverberation time because of the tiles, bath and other hard surface, whereas the bedroom is highly absorbent, and so has little reverberation.

Some reverberaton is needed to give body and fullness, especially for music. The ideal reverberation time depends on the type of music but lies between 1.75 and 2.5 seconds. Speech needs much less in the interests of clarity, although a little improves the subjective effect. For speech, between 0.5 and 0.75 seconds is about right. Broadcast studios intended for speech are designed around this value.

The reverberation time is not the same at all parts of an auditorium, nor is it the same for all frequencies. The decay curves are not linear, and the shape of these can vary. This is why the acoustics of concert halls differ so greatly. Some have long reverberation times at low frequencies, thus imparting a warm sonorous tone to the music, while others have longer treble times, giving a more brilliant effect. Usually a balance is sought, and modifications are sometimes made for different types of music by introducing absorbent panels and resonators at various points.

Most halls in which public address systems are installed have much longer reverberation times than is the optimum for speech, resulting in loss of clarity. With temporary installations little can be done other than angling column loudspeakers to avoid aiming the sound directly at a wall, as will be described in a later chapter.

With permanent systems some influence may be brought to bear on the hall administrators to improve acoustics by use of suitable absorbent materials. One of the most effective and visually acceptable is the provision of a heavy curtain across the rear wall of the platform. This could be drawn back for live musical events that require a longer reverberation time.

Wood-veneered ply or hardboard hung on wooden battens makes an excellent finish for the main auditorium walls. This absorbs low frequencies which are usually the most troublesome, yet does not deaden the higher register, as Table 3 shows. Putting fibreglass in the gap improves the low frequency absorption and also improves heat insulation, a useful bonus. These measures also assist in the prevention of feedback as described in Chapter 11.

The reverberation time of a hall can be calculated at different frequencies from the absorption coefficients and Sabine's formula:

$$T_r = \frac{0.161 \ V_m}{S_a} \quad \text{or} \quad T_r = \frac{0.05 \ V_f}{S_a}$$

In which T_r is the reverberation time in seconds; V_m is the volume in m^3; V_f is the volume in ft^3; S_a is the total absorption obtained by multiplying the individual areas of the material (m^2 for first equation, ft^2 for the second).

This formula gives results that are close to the measured ones except for near anechoic (acoustically dead) conditions.

3 Audio engineering principles

Having considered the principles of acoustics, we will now turn to those of electrical signal handling in the public address system. Some of those already dealt with in the previous chapter such as phase difference and addition apply also to electrical signals.

Much of the way signals behave is due to frequency discrimination in the circuits to which they are applied. This in turn depends mainly on two factors, capacitance and inductance. Their presence may be inadvertent, due to stray values in wiring and components, or it may be deliberate by the use of capacitors and inductors in filter circuits.

Capacitive reactance

Firstly, let us consider exactly what a basic capacitor is. It consists of two conductors having a large area in close proximity to each other. When a voltage source is connected across it, electrons are drawn from one surface and rush through the source to the other. There is thus a deficiency on the first and a surplus on the second. The capacitor is then said to be charged and if the source is disconnected, a voltage equal to that of the source can be measured across the capacitor.

Charging a capacitor is not like filling a bottle, in that once it is full it will accept no more. If the source voltage is increased, the charge will increase right to the point when the insulation between the surfaces breaks down. It is more like inflating a balloon which will take more and more until finally it bursts.

At the instant of connecting the capacitor to the source, the current flow is large as there is virtually no opposition to it. But as it becomes charged, its rising internal potential opposes that of the source, thus reducing the charging current. This current decreases until the internal potential and source voltage are equal, at which point it ceases. The charging curve is therefore not linear but exponential.

If removed from the source and applied to an external load, the capacitor will discharge in a similar manner, the current decreasing as the potential falls. Even without an external load a capacitor will discharge itself in time because of internal leakage through the imperfect insulation.

If a capacitor is partly charged, and then the source voltage is reversed, it will discharge and begin to charge to the opposite polarity. If the source is again reversed before the recharge is complete, it will discharge once more and start a further charge to its original polarity. This process can be kept going indefinitely, with current flowing in and out at each polarity reversal.

So, an a.c. source will keep current flowing, providing the reversals occur before the capacitor is fully charged. It should be noted that as the maximum current flows before the voltage starts to rise, maximum current and voltage do not occur at the same time. The current leads the voltage by 90°.

If the reversals are very rapid, charging and discharging will always take place at the start of the curve where the current is greatest. If reversals are slow, the capacitor will be well charged and the current be decreasing before the next reversal comes. So, the average charging current passing through a capacitor depends on how rapid the reversals are. In other words, the magnitude of the current depends on the frequency of the a.c. source – the higher the frequency, the greater the current.

It also depends on the size of the capacitor. One having a large capacitance takes longer to charge and discharge, and so will still be operating at the early high-current portion of its curve at the slower low frequencies. The larger the capacitance, the higher the current.

Both frequency and capacitance are therefore factors governing the current flow through a capacitor. They are combined in the property termed *capacitive reactance,* symbol X_c, which is to a capacitor what resistance is to a resistor, hence the unit is the ohm. The formula for calculating it is:

$$X_C = \frac{1}{2\pi fC}$$

where f = frequency in Hz, and C = capacitance in farads. For microfarads the formula becomes:

$$X_C = \frac{10^6}{2\pi fC}$$

The capacitor thus offers a high impedance to low-frequency signals and a low impedance to high frequencies, so enabling it to be used as an element in frequency-selective filter circuits. The effect is proportional, at half the frequency the reactance is double and the current is half. The response is thus said to roll off at 6 dB per octave.

Inductive reactance

When a current flows through a straight wire, a circular magnetic field surrounds it which is made up of individual lines of force. If the wire is wound in the form of a coil, the lines link up to produce a concentrated field in the form of loops passing longitudinally through the centre of the coil and around its exterior.

Whenever a magnetic line of force cuts across a conductor, it induces an electromotive force (e.m.f.) in it. This is the principle on which all electric generators depend. It matters not whether the conductor moves or the magnetic field, nor does it matter whether the field is produced by a permanent magnet or an electromagnet.

When the field produced by a coil builds up or collapses and so cuts across its own windings, an e.m.f. is induced in them. This always *opposes* the original voltage that produced the current, because its polarity is opposite to it. Thus the effective voltage acting in the circuit is that of the applied voltage minus the self-induced voltage.

In the case of a d.c. supply, the effect is momentary. When the current starts to flow, the opposing e.m.f. inhibits it so that it builds up slowly to its maximum. After this, the field is stationary and so has no effect. On removing the applied voltage the e.m.f. generated by the collapsing field tries to perpetuate the current. Quite high voltages can be induced by a sudden collapsing field.

With a.c., the field is constantly changing and so the current is continuously opposed by the back e.m.f. This is the property which is termed *inductive reactance*. Because the current builds up slowly it lags behind the applied voltage by 90°, so maximum current and voltage do not occur at the same time.

The opposing or back e.m.f. is proportional to the speed of the changing field, being highest when the speed is greatest. As the rate of change increases with frequency, it follows that the reactance is not constant but increases as the frequency rises.

It is also dependent on the inductance of the coil. This is a rather complex factor depending on the total number of turns, the turns per inch, length and diameter of the coil and the number of layers. Also affecting inductance is whether the coil is air-cored, iron-cored (increasing inductance) or brass-cored (decreasing inductance). The latter applies when brass slugs are used for tuning r.f. coils. The formula for inductive reactance is:

$$X_L = 2\pi f L$$

In which X_L is in ohms, f = frequency in Hz, L is inductance in henries.

Thus at low frequencies the inductor offers minimum impedance, but at high frequencies the impedance is also high. Along with the capacitor which has the opposite effect, it can be used to filter audio signals. Like the capacitor, the reactance is proportional, doubling the frequency doubles the reactance and thereby halves the current. The reactance thus rolls off at 6 dB per octave but in the opposite direction to the capacitor. (See reactance chart, Figure 17.)

Time constant

When a resistor is connected in series with a capacitor the current flow is restricted and so the capacitor takes longer to charge. The frequency response of the combination is thus affected and is governed by both the value of the capacitance and the resistance. Such combinations are frequently used for simple 6 dB/octave roll-off circuits as the frequency can easily be altered by changing the value of the resistance.

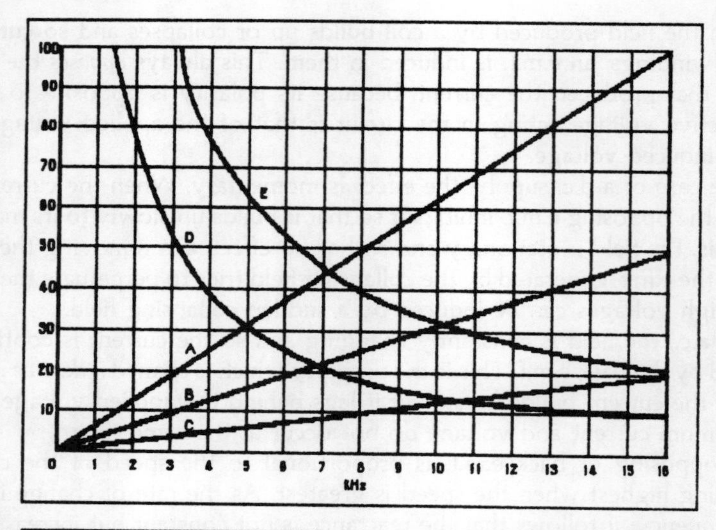

INDUCTIVE REACTANCE

Inductance (Henries)	Curve	Multiply Scale
0.0002	C	× 1
0.0005	B	× 1
0.001	A	× 1
0.002	C	× 10
0.005	B	× 10
0.01	A	× 10
0.02	C	× 100
0.05	B	× 100
0.1	A	× 100
0.2	C	× 1000
0.5	B	× 1000
1.0	A	× 1000

CAPACITIVE REACTANCE

Capacitance (μF)	Curve	Multiply Scale
0.0005	E	× 1000
0.001	D	× 1000
0.005	E	× 100
0.01	D	× 100
0.05	E	× 10
0.1	D	× 10
0.5	E	× 1
1.0	D	× 1
5.0	E	÷ 10
10.0	D	÷ 10
50.0	E	÷ 100

Figure 17 Chart of inductive and capacitive reactances.

As the charging current decreases exponentially with time, it decreases by very small amounts near maximum. It is thus difficult to determine the exact time when the capacitor is actually fully charged. The time constant of an *RC* combination is therefore considered to be the time at which the voltage across the capacitor reaches 63% of maximum, as this makes a straightforward relationship between time in seconds, capacitance, and resistance.

Although much less used than the *CR* circuit, a time constant is also applicable to a series inductor/resistor circuit. It is defined as the time when the current through an inductor reaches 63% of its maximum. The respective formulae are

$$t = CR \qquad t = \frac{L}{R}$$

in which t is the time in seconds.

Impedance

Impedance is the total opposition offered to an alternating current resulting from the capacitive reactance, inductive reactance, and resistance. As the current leads the voltage with capacitance, lags with inductance, and is in phase for resistance, the phase angle in any circuit depends on the respective values of the three factors, or two if only two are present.

The impedance formulae for various combinations and the phase angles are as follows:

Series capacitance/resistance: $Z = \sqrt{(R^2 + X_C^2)}$

phase angle: $\phi = \tan^{-1} \dfrac{X_C}{R}$

Series inductance/resistance: $Z = \sqrt{(R^2 + X_L^2)}$

phase angle: $\phi = \tan^{-1} \dfrac{X_L}{R}$

Series inductance/capacitance/resistance: $Z = \sqrt{[R^2 + (X_L - X_C)^2]}$

phase angle: $\phi = \tan^{-1} \dfrac{(X_L - X_C)}{R}$

Parallel capacitance/resistance: $Z = \dfrac{RX_C}{\sqrt{(R_2 + X_C^2)}}$

phase angle: $\phi = \tan^{-1} \dfrac{R}{X_C}$

Parallel inductance/resistance: $Z = \dfrac{RX_L}{\sqrt{(R^2 + X_L^2)}}$

phase angle: $\phi = \tan^{-1} \dfrac{R}{X_L}$

Series inductance/resistance, parallel capacitance:

$$Z = \frac{1}{\sqrt{\left(\frac{R}{R^2 + X_L^2}\right)^2 + \left(\frac{X_L}{R^2 + X_L^2 - X_C}\right)^2}}$$

$$\text{phase angle:} \quad \phi = \tan^{-1} \frac{\left(\frac{X_L}{R^2 + X_L^2} - X_C\right)}{\left(\frac{R}{R^2 + X_L^2}\right)}$$

Resonance

When capacitance and inductance are in series, there is one frequency at which both reactances are equal. However, as one produces leading and the other lagging current, they cancel to produce zero reactance and the only impedance in the circuit is that of the resistance. This is known as the resonant frequency f_r. The formulae are:

$$X_L = X_C; \qquad f_r = \frac{1}{2\pi\sqrt{(LC)}}$$

$$\text{Impedance at resonance:} \quad Z = R$$

With parallel circuits, at resonance when there is zero phase difference and reactance, the capacitive and inductive reactances are not exactly equal. Resistance affects the resonant frequency:

$$f_r = \frac{1}{2\pi}\sqrt{\left(\frac{1}{LC} - \frac{R^2}{L^2}\right)}$$

$$\text{Impedance at resonance:} \quad Z = \frac{L}{CR}$$

Q factor

Q defines the quality of a component or circuit as

$$\frac{2\pi \times \text{maximum energy stored in 1 cycle}}{\text{energy dissipated in 1 cycle}}$$

For an inductor, the energy stored $= LI^2$ joules; while the energy dissipated $= \frac{I^2R}{f}$ joules

Hence

$$Q = \frac{2\pi LI^2 f}{I^2 R} \quad \text{or} \quad \frac{2\pi fL}{R} \qquad \text{which is} \quad \frac{X_L}{R}$$

So the Q of an inductor is simply its reactance divided by its resistance or any resistance in series with it.

Q thus increases with frequency until skin effect increases R, causing Q to rise less rapidly. Then the self-capacitance reduces X_L and also Q, until it is completely cancelled and Q falls to zero. This occurs at the self-resonant frequency of the inductor.

In the case of a capacitor

$$Q = \frac{1}{2\pi fCR} \quad \text{or} \quad \frac{1}{X_C R}$$

Q at resonance

When both inductance and capacitance are present in a series circuit, Q is as follows:

$$Q = \frac{1}{R}\sqrt{\left(\frac{L}{C}\right)}$$

It is a paradox that the voltage across either the capacitor or inductor at resonance is *greater* than the applied voltage when it may be expected to be only a half in a series circuit. It is in fact Q times the applied voltage, hence Q is also called the magnification factor.

For a resonant parallel circuit containing inductance and capacitance, assuming negligible loss in the capacitor, Q is equal to that of the inductor. With a Q higher than 10, the current in each branch is Q times the supply current, so in this case Q is a *current* magnification factor.

The bandwidth at the -3 dB points either side of the resonant frequency is inversely proportional to Q and is given by

$$\frac{f_r}{Q}$$

Gyrator

An inductance can be simulated by electronically reversing the action of a capacitor. This is useful as large inductors to tune to low frequencies are expensive, and prone to pick up hum. The device known as a gyrator consists of an inverting and non-inverting amplifier stage with the output of each connected to the input of the other. An impedance connected at one produces an 'image' at the other equal to

$$Z_2 = \frac{1}{Z_1 g_1 g_2}$$

in which Z_2 is the image impedance, Z_1 is the actual impedance, g_1 and g_2 are the slopes of the two amplifiers in amperes/volt. These slopes are also known as the *gyrator constants*.

If the actual impedance is a pure capacitance of which the reactance is

$$\frac{1}{2\pi fC}$$

then the image impedance becomes:

$$Z_2 = \frac{1}{\frac{1}{2\pi fC} g_1 g_2} \quad \text{or} \quad = \frac{2\pi fC}{g_1 g_2}$$

This is the reciprocal of the actual capacitive reactance and analogous to the expression for inductance: $2\pi fL$. So the image has the characteristics of a pure inductance of which the value in henries is equal to the value of the actual impedance in farads divided by the product of the gyration constants. It is noteworthy from this that the image inductance is greater when the amplifier slopes are lower, but the Q is lower.

The image inductance can be combined with a real capacitor to form a resonant circuit. Providing the g of both amplifiers is the same, and the two capacitors have the same value, the resonant frequency is

$$f_r = \frac{g}{2\pi C}$$

and, providing both amplifier input and output impedances are the same, with R being the value of the input impedance of one amplifier in parallel with the output impedance of the other, Q is given by

$$Q = \frac{gR}{2}$$

An asymmetrical gyrator can be made from a single transistor as the inverting amplifier, and a resistor R_b across the collector to base as the 'non-inverting amplifier'. The Q is very low but the circuit is suitable as a supply-line series hum filter. The image impedance is

$$Z_2 = \frac{R_b}{g_1 Z_1}$$

Filters

Most filters are a combination of capacitors and inductors to give various characteristics and roll-off values. Their behaviour is determined by the terminating impedance, so it is designed for a particular value which is termed the *design impedance*. When so designed, the impedance measured at the input terminals is the same as that of the output and is called the *iterative impedance*. Filters may then be cascaded providing the last one is terminated with the correct impedance.

Constant-*k* low-pass filters (Figure 18)

These can be either what are called T-type or π-type, this designation resulting from their appearance when drawn in a circuit. They have three components that give a nominal 18 dB/octave attenuation above the cut-off frequency which is -3 dB below the pass-band level. Both types have the same properties, but as the π-type has only one inductor with two capacitors against the two inductors and one capacitor of the T-type, it is generally preferred. With the T-type the calculated inductance is *halved* for each of the values of the two inductors, and the capacitance value is halved for the two capacitors of the π-type to give the values of the two capacitors. The same formulae applies to both types:

$$Z = \sqrt{\frac{L}{C}} \quad f = \frac{1}{\pi\sqrt{(LC)}} \quad L = \frac{Z}{\pi f} \quad C = \frac{1}{\pi f Z}$$

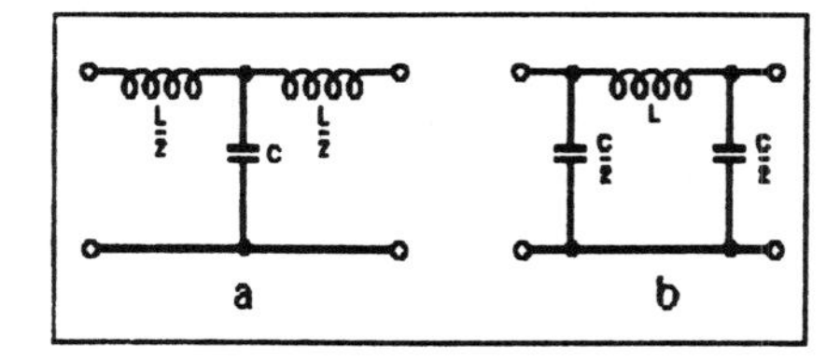

Figure 18 (a) T-type low-pass; (b) π-type low-pass.

in which L is the inductance in henries; C is the capacitance in farads; f is the cut-off frequency; and Z is the design impedance.

Constant-*k* high-pass filters (Figure 19)

Either T-type or π-type filters can be used to provide the nominal 18 dB/octave attenuation below the cut-off frequency. Both have the same properties, but in this case the T-type has only one inductor whereas the π-type has two. Choice may be influenced by the fact that the π-type has a d.c. path across input whereas the T-type has not. The calculated capacitance is *doubled* for each of the capacitors of the T-type filter, while the inductance is *doubled* for each inductor of the π-type. The formulae for both is

$$Z = \sqrt{\frac{L}{C}} \quad f = \frac{1}{4\pi\sqrt{(LC)}} \quad L = \frac{Z}{4\pi f} \quad C = \frac{1}{4\pi f Z}$$

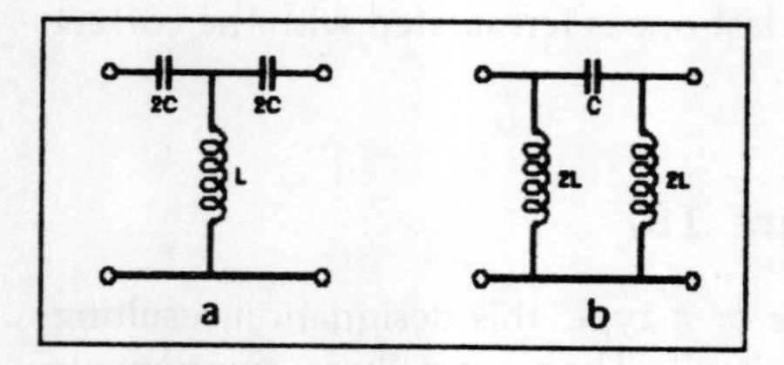

Figure 19 (a) T-type high-pass; (b) π-type high-pass

m-Derived filters

These include a tuned circuit which gives a very high attenuation at the resonant frequency, but a much lower attenuation than a constant-*k* filter at frequencies, beyond resonance. The quantity M is a function of the ratio between the cut-off frequency f_c and the resonant frequency f_r. It is

$$m = \sqrt{\left[1 - \left(\frac{f_c}{f_r}\right)^2\right]}$$

An m value of 1.0 is equivalent to a constant-*k* filter. As it gets lower, attenuation at the cut-off frequency gets steeper, but that above it for low-pass filters, and below it for high-pass filters, becomes less.

Table 5 Attenuation related to m values

m values:	0.4	0.5	0.6	0.7	0.8	0.9
dB attenuation	7.6	10	12	15	19	26

Attenuation beyond cut-off drops to the above levels

T-type *m*-derived low-pass filter (Figure 20a)

The circuit is the same as for a constant-*k* filter, but a third inductor is added in series with the capacitor. The values are calculated for constant *k* then related to *m* as follows:

$$L = \frac{Z}{\pi f} \quad C = \frac{1}{\pi f Z} \quad l_1 = l_2 = \frac{mL}{2} \quad l_3 = \frac{(1 - m^2)}{4m} L \quad c = mC$$

π-Type m-derived low-pass filter (Figure 20b)

A third capacitor is shunted across the series inductor of the constant-*k* circuit. Having one inductor to the T-type's three makes it an obvious choice. There is no d.c. path in either. Values are derived from the constant-*k* formulae:

$$L = \frac{Z}{\pi f} \quad C = \frac{1}{\pi f Z} \quad c_1 = c_2 = \frac{mC}{2} \quad c_3 = \frac{(1 - m^2)}{4m} C \quad l = mL$$

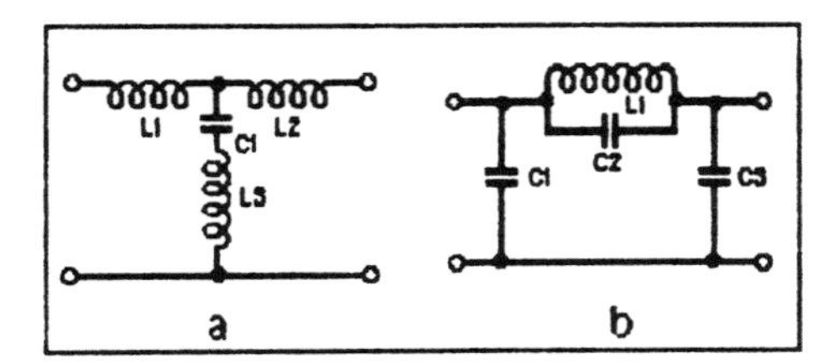

Figure 20 (a) T-type m-derived low-pass; (b) π-type m-derived low-pass.

T-type *m*-derived high-pass filter (Figure 21a)

A third capacitor is added in series with the inductor of the constant-*k* circuit. Then *m* modified values are:

$$L = \frac{Z}{4\pi f} \quad C = \frac{1}{4\pi f Z} \quad c_1 = c_2 = \frac{2C}{m} \quad c_3 = \frac{4m}{1 - m^2} C \quad l = \frac{L}{m}$$

π-Type *m*-derived high-pass filter (Figure 21b)

The series capacitor in the equivalent constant-*k* circuit is shunted by a third inductor. The values are:

$$L = \frac{Z}{4\pi f} \quad C = \frac{1}{4\pi f Z} \quad l_1 = l_2 = \frac{2L}{m} \quad l_3 = \frac{4m}{1 - m^2} L \quad c = C$$

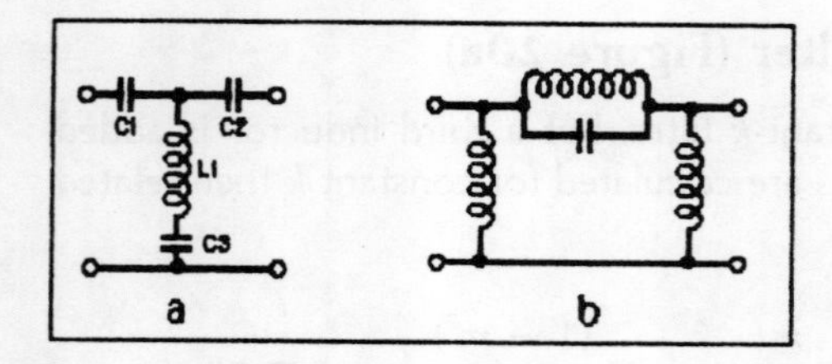

Figure 21 (a) T-type m-*derived high-pass; (b) π-type* m-*derived high-pass.*

Adding two a.c. voltages or currents

This can be done geometrically by drawing vectors to scale and measuring the resultants as described in the previous chapter, or by the following formula:

$$V = \sqrt{(\cos\theta \times 2v_1v_2 + v_1^2 + v_2^2)}$$

in which θ is the angle of phase difference, v_1 is the first voltage or current, v_2 is the second voltage or current.

Impedance matching

If maximum power is to be transferred from one circuit to another their respective impedances must be the same. A high-impedance circuit has a high voltage with low current whereas a low-impedance circuit has low voltage and high current. It follows that if a low-impedance circuit is connected to a high-impedance source, the high voltage will drive an excessive current through the circuit which will either damage the load or source, or greatly reduce the voltage available.

On the other hand, connecting high-impedance load across a low-impedance source results in low current and low power transfer.

In the case of a loudspeaker system, for maximum power transfer, the impedance of the system should equal that of the amplifier output impedance or be a little higher. In other cases where maximum voltage rather than power transfer is the main consideration, the load impedance should be higher than the source.

The reason for this becomes evident if we consider the circuit in Figure 22. The source can be drawn as a generator with series resistor to represent the source impedance. The load is a resistor of equal impedance. Thus the two resistors are effectively in series across the generator and behave as a potential divider. So, with the load and source impedances equal, the voltage appearing across AB, is only half the generator output, and the coupling is just 50% efficient.

If the load is ten times the value of the source impedance, the signal appearing across AB becomes 10/11ths of the generator output and the coupling is over 90% efficient. Power transfer though is greatly reduced.

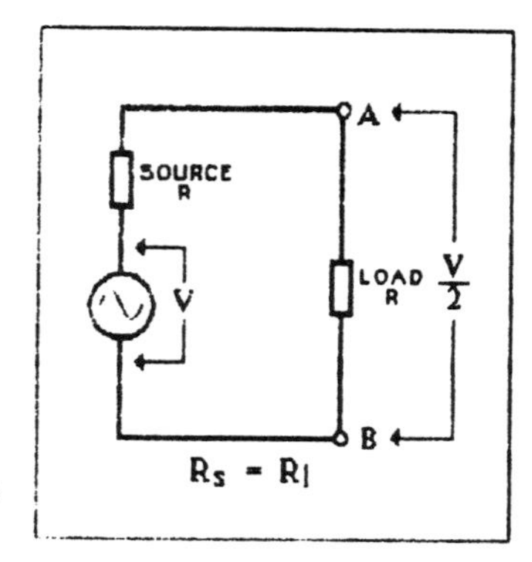

Figure 22 When source and load impedances are equal output voltage is halved.

It is clear that the type of coupling depends on whether power is to be transferred such as with loudspeaker systems, or whether voltage is the important factor as with most input circuits.

Matching transformers

The magnetic field produced by the primary winding of a transformer cuts across the windings of the secondary, which is usually overwound on the primary. An emf is thereby induced in it which is proportional to the turns ratio between the windings, while the current available is inversely proportional. The product of voltage and current in the secondary is thereby the same, apart from minor losses, as that in the primary.

This is a familiar principle when dealing with power supply circuits, but it follows that the impedance of the secondary circuit is also different from that of the primary, and likewise depends on the turns ratio. A secondary giving higher voltage and lower current than the primary has a higher impedance, and vice versa.

The transformer thus provides a convenient method of impedance matching and can be used for microphone input circuits and loudspeaker systems. Public address amplifiers have internal output transformers to match the output stage to 100 V or 70 V loudspeaker lines.

The turns ratio between primary and secondary windings is the square root of the ratio of the impedances. Thus:

$$tr = \sqrt{\frac{Z_1}{Z_2}}$$

Resistor networks are used for matching a circuit of one impedance to that of another, and also for attenuation. The main types are: L, T, and π, so called because the circuit configuration resembles those letters. Each type except L can be *symmetrical, asymmetrical, balanced* or *unbalanced.*

L-type network

This is used for matching two dissimilar impedances with the minimum attenuation. The high (Z_H) must be connected to terminals 1 and 2, and the low (Z_L) to 3 and 4 (Figure 23a), irrespective of which is the source and which is the load. The values are calculated by

$$R_1 = \frac{Z_H Z_L}{R_2} \qquad R_2 = \sqrt{\left(\frac{Z_H Z_L^2}{Z_H - Z_L}\right)}$$

The insertion loss in dB is

$$10 \log_{10} \left(\frac{Z_L + R_2}{R_2}\right)^2 \frac{Z_H}{Z_L}$$

A *balanced* version of the *L* network (Figure 23b) can be produced by splitting the value of R_1 into two equal parts and putting them in each series leg.

T-type network

The symmetrical T-type pad (Figure 23c) is used between two circuits of the same impedance (Z) to introduce attenuation without upsetting the existing

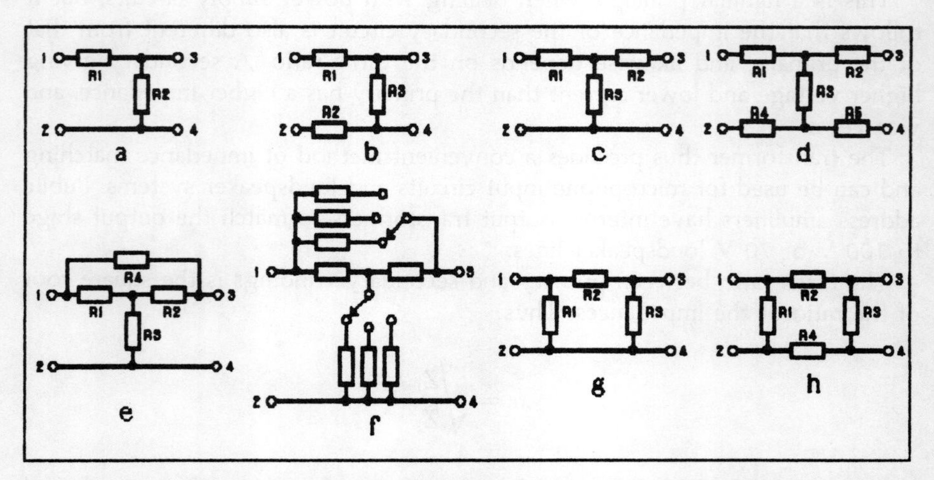

Figure 23 (a) Unbalanced L-network; (b) balanced L-type;(c) unbalanced T; (d) balanced T-type; (e) bridged T-type; (f) three-level attenuator using bridged T-pads; (g) unbalanced π-type; (h) balanced π-type.

match. The design is simplified if the required attenuation is defined as a current ratio (N) rather than in decibels. Then:

$$R_1 = R_2 = \left(\frac{N-1}{N+1}\right) \qquad R_3 = Z\left(\frac{2N}{N^2-1}\right)$$

Asymmetrical T-type network

The asymmetrical T-pad (Figure 23c) is used to insert a required amount of attenuation (N) between a high impedance (Z_H) and a low impedance (Z_L) circuit. There is a minimum loss, so attenuation cannot be lower than this. It is governed by the ratio of the two impedances, so $N > Z_H/Z_L$. The values are:

$$R_3 = \sqrt{(Z_H Z_L)}\left(\frac{2N}{N^2-1}\right) \qquad R_1 = Z_H\left(\frac{N^2+1}{N^2-1}\right) - R_3$$

$$R_2 = Z_L\left(\frac{N^2+1}{N^2-1}\right) - R_3$$

The high impedance Z_H must be connected across terminals 1 and 2, while the low impedance Z_L is connected across 3 and 4.

The network can be balanced (Figure 23d) by halving the values of R_1 and R_2, and adding R_4 and R_5 with values $R_1 = R_4$, and $R_2 = R_5$.

Bridged T-network

This pad is used for switched attenuators where several values of attenuation are required. Although an extra resistor is used for bridging each network, two are common to all, so for a three-level attenuator eight resistors are used instead of nine. By bridging R_1 and R_2 of a symmetrical T-pad with R_4, the attenuation can be changed by altering only two resistors, R_3 and R_4, instead of three. Thus the switch need have only two poles. The values are:

$$R_1 = R_2 = Z \quad R_3 = \frac{Z}{(N-1)} \quad R_4 = Z(N-1)$$

π-Type network

A π-type pad has the same properties as the T-pad and can be used as an alternative, having the same number of resistors. In some cases the values of one type may be closer to preferred values than those of the other. The shunt resistance of both types are the same, but the series value is higher with the π-type, which may be a consideration if d.c. is involved. Values for the symmetrical type are:

$$R_1 = R_3 = Z\left(\frac{N+1}{N-1}\right) \qquad R_2 = Z\left(\frac{N^2-1}{2N}\right)$$

Asymmetrical π-type network

Values for this are more complex and are most easily calculated by first designing an asymetrical T-pad, then using its values for the π-pad, as follows:

$$R_1 = \frac{R}{TR_2} \quad R_2 = \frac{R}{TR_3} \quad R_3 = \frac{R}{TR_1}$$

in which *TR* are the T-pad values; and

$$R = TR_1TR_2 + TR_1TR_3 + TR_2TR_3$$

Balancing is achieved by halving the value of R_2 and adding R_4 of the same value. Thus only four resistors are needed instead of five for the equivalent balanced T-Type network.

Mechanical resonance

All physical objects will resonate or vibrate more strongly at one particular frequency and its multiples than any other. The frequency is dependent on the mass of the object and its springiness or that of the surrounding medium. Mass incidentally is not quite the same as weight, it denotes the quantity of matter which is fixed, whereas the weight varies with gravity, which in turn varies with location.

Its particular application in public address work is with the moving parts of microphones and loudspeakers. When a wide range of frequencies is applied, the motion increases at the resonant frequency. This results in higher electrical output at that frequency in the case of the microphone, and higher sound output from the loudspeaker.

The frequency response is thus not flat but has a number of peaks. These change the character of the sound, but more seriously for public address work, they initiate feedback.

Fourier analysis

Any continuous periodic waveform can be analysed to produce a succession of pure sine waves consisting of a fundamental and harmonics. The analysis involves rather complicated mathematics and is not usually required for public address work. There is one aspect though that does have an application.

A square wave is no exception to the rule, and can be analysed as consisting of a sine wave fundamental with an infinite number of odd harmonics of decreasing amplitude. It follows from this that an amplifier must have a level and extended frequency response to handle square waves. If it is deficient, the shape of the square as seen on an oscilloscope will be distorted.

The use of square waves which are generated easily is thus a convenient means of checking the response of an amplifier at one glance, without having to run a whole series of measurements at different frequencies.

Another situation where the principle is encountered is when an amplifier is overloaded. The result is usually the clipping of the peaks of reproduced waveforms. The flat tops are a virtual square waveform and so a large number of odd harmonics are thereby generated. It is these spurious harmonics that make overload distortion so unpleasant, rather than the missing wave tops.

Electromagnetic devices

We have already touched on magnetic behaviour in our consideration of inductance and transformer matching. Other applications of magnetic principles that are commonly encountered in public-address work are: microphones, loudspeakers, analogue meters, relays, tape recording, induction loops, standby generators etc. A basic knowledge of magnetic theory will thus be helpful in working with these and is here presented.

There is often confusion over the terms used to denote various magnetic properties. This has not been helped by the changing of some of the units, from cgs (centimetre, gramme, second) to the clumsier MKS (metre, kilogramme, second), resulting in both old and new terms being encountered.

Magnetic moments

A magnetic field is produced when an electric current, which consists of electrons, flows in a circuit. The motion of orbiting electrons in the atom thereby generates a magnetic field, with further fields being produced by the electron spin. These are termed the *magnetic moments,* and the moment of the whole atom is the vector sum of the individual ones.

The sum may be zero, in which case the material is termed *diamagnetic,* being non-magnetic. If there is a remaining moment, the material is said to be *paramagnetic,* and can exhibit magnetic properties.

Magnetization

As opposite magnetic poles attract each other, the molecules of a paramagnetic substance arrange themselves in circles so forming complete magnetic circuits. Thus the fields are all contained within the material and none appears outside.

When such a material is placed in a magnetic field, the internal fields line up under its influence and produce an external field which reinforces the applied one. If when the applied field is removed, the molecules revert to their former state, the induced field disappears, and the material is said to be magnetically *soft*. If some remain in line, these continue to exhibit an external field, and material is termed magnetically *hard*. The magnet so produced is called a *permanent magnet*. Magnet materials that are physically hard are usually magnetically hard, while those that are soft are generally magnetically soft too, but not always.

Magnetic poles

The lines of force emerge at or near the ends of the magnet. These are called the *poles*, and any specially engineered terminations to direct or concentrate the field are described as *pole-pieces*. The poles are commonly termed north and south respectively, but more accurately they are *north-seeking* and *south-seeking*. That is they are attracted to the Earth's magnetic poles that are at its geographic north and south. As opposite poles attract, the Earth's geographic north pole is the approximate location of its magnetic *south* pole. Furthermore, the magnetic and geographic poles do not coincide, which is why compass needles in England point to 16°W of true north, or nearer to NNW.

As a bar magnet is a string of molecular magnets end to end, the magnet can be broken across its centre to form two magnets, and further subdivided, each division producing magnets with a north and south pole. Two magnets can be joined end-to-end with a north and south pole in conjunction to become a single magnet. By dividing or joining, the pole strength is neither decreased or increased—the same number of lines of force run through the material.

Unit pole

A *unit pole* is defined as having the strength to exercise a force of 1 dyne (10^{-3} newtons) on a similar pole at a distance of 1 cm.

Flux

The symbol Φ represents the magnetic field, which consists of lines of force or flux each of which forms a continuous loop that passes externally from the north pole to the south, then internally back to the north pole. The cgs unit is the *maxwell* which denotes one line of force; the MKS unit is the *weber* which is equivalent to 10^8 maxwells. One weber produces one volt when reduced to zero at a uniform rate, across a coil of one turn in one second.

Lines of force take the path of least magnetic resistance (*reluctance*), and so choose an external path through a material having the least reluctance. If there is no choice and the surrounding material is homogeneous, as air is, for example, they take the shortest path. However, they also repel each other and can never cross, so they balloon out from the magnet, each following the shortest path that does not encroach on another line. The stronger the field, the closer and more dense they are.

Field strength

The symbol H represents the field strength, which is rated in air (strictly a vacuum) being the number of lines of force in a given area. The cgs unit is the *oersted,* which is one line of flux per cm^2. The MKS unit is the *ampere metre,* (1 kA/m = 12.5 oersteds). It applies in particular to an *applied field* from a magnetic source to distinguish it from the *induced field* in a magnetic substance resulting from that field.

Flux density

The symbol B represents the density of flux induced into a material by an applied field. It is defined by the number of lines of force passing through a surface of given area perpendicular to the direction of the lines. The cgs unit is the *gauss,* which is the flux produced by a field of one oersted, that is one line of flux per cm^2. The MKS unit is the *tesla,* which is 1 weber/m^2.

For a field in air,

$$B = H$$

Permeability

The symbol μ represents permeability. An applied field induces a greater field than itself in a paramagnetic material. The magnitude of the induced field is a product of the magnetizing force (applied field) and the permeability of the material. For diamagnetic substances the permeability is less than 1, for air it is 1, for paramagnetic materials it is greater than 1. For ferrous metals it can be several thousand. So:

$$\mu = \frac{B}{H} \quad \text{or} \quad B = \mu H$$

Reluctance

Reluctance is the magnetic resistance offered by a material having a length L, cross-sectional area a, and permeability μ:

$$r = \frac{L}{\mu a}$$

Saturation

A magnetizing force causes the internal magnetic particles in a material to line up and so produce a coherent external field. The stronger the force, the greater the number of internal magnets that line up until finally all are so aligned. The material is then magnetically saturated; no further increase in force will induce an increase in flux.

Remanence/retentivity

When a magnetizing force is removed, the induced field in a material drops. The remaining level is termed the remanence.

Coercivity

The symbol H_c denotes coercivity, which is the amount of applied reverse magnetic force needed to coerce the remaining magnetism of a material to zero after is has been magnetized to saturation. The material never ceases to be magnetic, rather the internal magnets are reorientated so that there is no coherent external field.

Electromagnetic induction

When the lines of force of a magnetic field cuts across a conductor, an electromotive force is induced. The magnitude of this is proportional to the rate of change of the flux. This is known as *Faraday's law*.

The current resulting from that e.m.f. sets up a magnetic field which opposes the original field. If that were produced by current in an adjacent conductor, the new field would produce an e.m.f. and a current in that conductor, which would oppose the original current flowing in it. If the field were produced by a moving magnet, its effect would be to oppose the motion of the magnet. So in all cases, the original action is opposed. This effect is described in *Lenz's law*.

The direction of the flux around a conductor can be determined by *Maxwell's corkscrew rule*. If a corkscrew is visualized as proceeding clockwise away from the observer, the direction represents the flow of the current from positive to negative, and the rotation, the direction of the flux. It should be noted here that the convention is to depict current flowing from positive to negative whereas it actually flows the other way, as free negative electrons flow toward the positive pole which is deficient in electrons.

When a conductor is formed into a coil, the lines of flux from adjacent turns link up to form loops passing along the centre of the coil and around the outside. The field so produced is the product of the current flowing and the number of turns, but not the applied voltage.

4 Microphones for public address

There is a considerable number of types and varieties of microphones, but very few are suitable for public address work. Many of the problems that arise are due to using unsuitable types. There are two basic requirements: first, the instrument should be directional, rejecting sound from the sides and rear. The second is that the frequency response should be as flat as possible. Both of these are essential for the reduction of feedback, and we will examine each more closely.

There are five main directivity patterns or polar responses, these being represented by a line plotted on a graph of concentric circles.

Omnidirectional

This type of microphone consists of a cone or diaphragm that has a sealed chamber behind it. The pressure in the trapped air is constant, so pressure variations on the front of the diaphragm push or pull it from its rest position. It thus responds to all pressure waves, irrespective of the direction of propagation, and so is called omnidirectional (Figure 24.)

Diffraction of rear pressure waves decreases as the wavelength approaches the diameter of the diaphragm, so the off-axis response diminishes slightly as the frequency increases. The smaller the diaphragm the higher the frequency at which the fall-off starts.

With waves coming on-axis, particle velocity of the air adds to the normal pressure wave, so increasing the response. The omnidirectional microphone is thus not truly omnidirectional but has a slightly greater response at the front for all frequencies, with a decreasing response at the rear as the frequency rises (Figure 25a). However, for most practical purposes it can be considered omnidirectional, and its lack of rejection of non-frontal sounds makes it quite unsuitable for public address work.

Cardioid

If the back surface of a diaphragm is vented to the atmosphere, pressure waves affect both surfaces equally, and there is zero movement. When an acoustic resistance is placed in the air vent it reduces pressure to the back surface, so a frontal wave assisted by direct particle velocity exerts greater force on the front and produces movement.

When a wave arrives from the rear, pressure is still greater at the front because of the unrestricted access, but the direct particle velocity now affects the back to a limited extent through the vent, as well as reduced wave pressure. So the combined forces on the back and front are approximately equal and little diaphragm motion occurs.

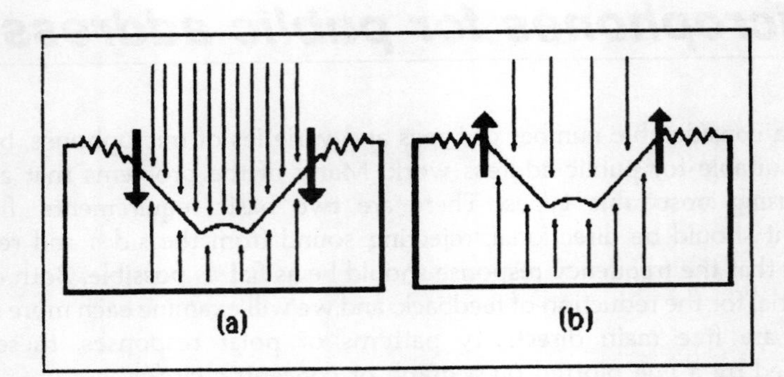

Figure 24 (a) High external air pressure pushes diaphragm inward; (b) low pressure allows internal pressure to push it outward.

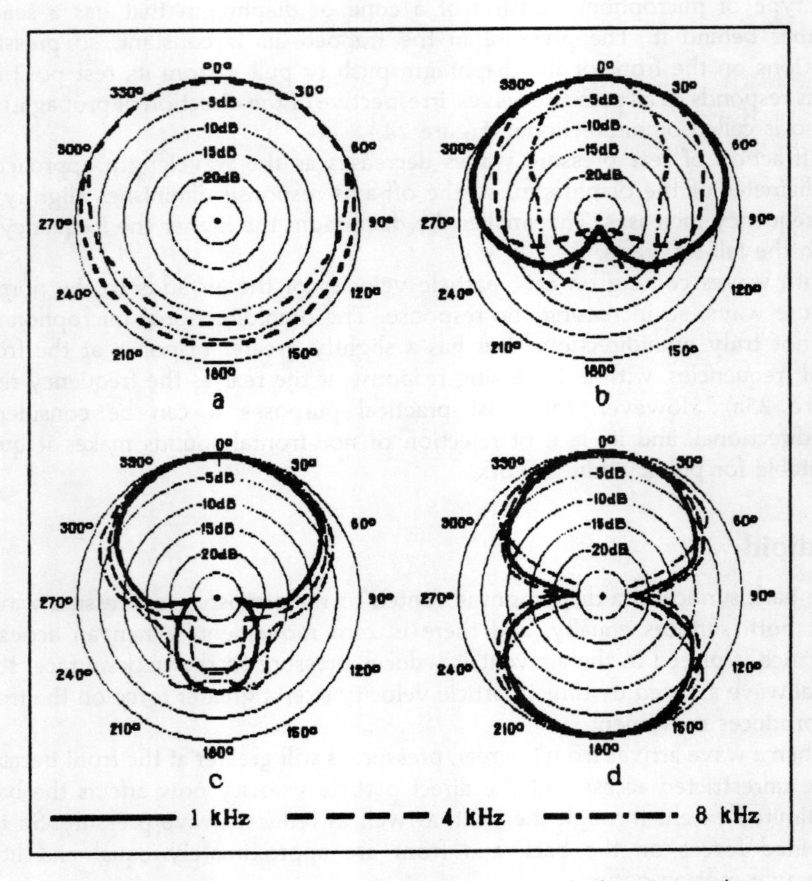

Figure 25 (a) omnidirectional; (b) cardioid; (c) hypercardioid; (c) figure-eight.

Waves arriving from the side impart no direct velocity to either front or back, but the acoustic resistance in the rear passage reduces the pressure at the back causing some movement of the diaphragm, though it is not as great as for a front-arriving wave.

The principle is termed pressure gradient, because there is a varying gradient in pressure between the front and back of the diaphragm depending on the direction of the sound source (Figure 26). The polar response seen at Figure 25b is heart-shaped, hence is called cardioid. The response is 20–30 dB down at the rear and some 6 dB down at the sides.

Directivity is greater at higher frequencies because high-frequency waves from the front are not diffracted to the side vents and so do not reach the back of the diaphragm. The change of output with angle of incidence of arriving sound wave is:

$$\frac{1 + \cos\theta}{2}$$

The rear rejection makes it more suitable for public address work than the omnidirectional microphone, as feedback from the auditorium is thereby reduced. However, the low-side rejection limits its usefulness as much of the unwanted pickup of reflected sound comes from the side walls.

Super/hypercardioid

These operate on the same principle as the cardioid but the acoustic resistance is reduced and vent size modified. This increases the pressure at the back of the diaphragm, resulting in cancellation of side-propagated waves. The result is a

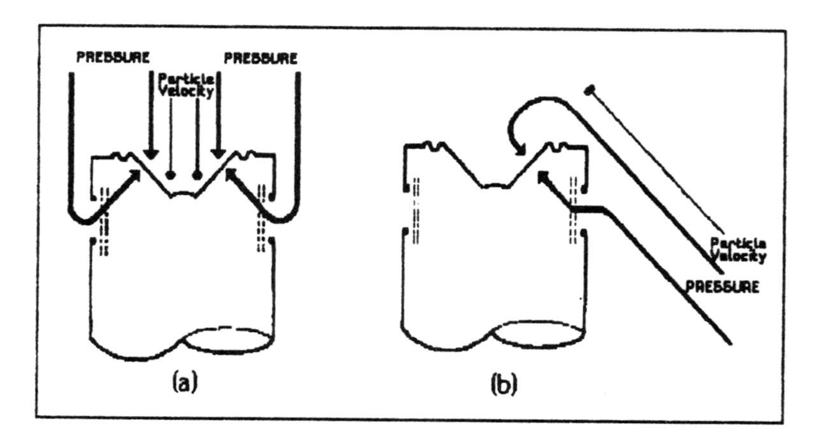

Figure 26 (a) Pressure on both sides of the diaphragm partly cancels, but front pressure is augmented by a particle velocity; (b) pressure from a rear wave cancels but there is no particle velocity reinforcement.

narrower forward lobe with side response reduced to −8.7 dB. This is called the *supercardioid* and it has two maximum rejection points at 125° off-axis.

If the resistance is decreased further, the side response is reduced still more to around −10 to 12 dB, but the effect of particle velocity from rear-arriving sound on the back of the diaphragm now overcomes the pressure at the front. As a result, a small lobe appears at the rear of the response and the device becomes a *hypercardioid* (Figure 25c). It has its maximum rejection at two 110° off-axis points.

Directivity

In spite of the small lobe at the rear of the hypercardioid where the cardioid has none, the reduced sensitivity at all sides gives a greater overall rejection of non-frontal sounds than that of the cardioid.

If an omnidirectional and a cardioid or hypercardioid microphone of equal sensitivity are placed facing a sound source in a reverberant sound field, the acoustic power received by the omnidirectional microphone is greater than that of the other microphone because it receives reflected sound from all directions. The ratio of this power compared to that received by the omni is a measure of the total directivity of that microphone and is described as the *directivity coefficient* or the *sound power concentration*.

Since the sound intensity decreases according to the square of the distance, a directional microphone can be further away from the sound source than an omnidirectional one by a distance equal to the square root of the directivity coefficient, for the same amount of ambient noise or feedback.

A cardioid has a coefficient of 3, and so can be placed $\sqrt{3} = 1.73$ times the distance from the source as the omnidirectional for a given feedback level. The hypercardioid with a coefficient of 4 can be placed $\sqrt{4}$ or twice as far.

Because reflections which cause feedback come from all directions it is the overall rejection which is important. The hypercardioid rejects the most and so is best for reducing feedback. The difference in the above figures may seem small, but the difference in feedback is quite noticeable when working in difficult acoustic conditions.

Particularly when the cardioid is used with its side towards the auditorium such as for interviews, the low 6 dB side rejection can produce feedback, poor intelligibility due to a high degree of amplified reverberation, or both.

If data may not be available for a particular model, it is usually possible to distinguish between an omnidirectional and a cardioid or hypercardioid, even though the style of microphone housings differ considerably. The omnidirectional has no vents in the side, whereas the others have some. Distinguishing a cardioid from a hypercardioid is less certain. The vents in the latter are often larger than the cardioid but this can vary with different models.

Ball-ended microphones are usually cardioids or hypercardioids, the rear air access being through the base of the ball.

Figure eight

A response which looks like a figure eight is obtained from the ribbon type of transducer when both front and back are exposed to the air, the sides being masked by the magnet poles (Figure 25d). Pressure is the same on both sides of the ribbon, so it is actuated not by pressure differences but by particle velocity. It is thus often called a velocity microphone.

Pads or other modifications are often made to reduce the rear response to a small rear lobe, then the microphone is close to being a hypercardioid. In this form it is an excellent choice for public address work, as ribbons inherently have a flat response.

Gun/interference tube

These microphones, which are often used for outdoor film sound recording, consists of a long tube having a row of holes or slots along one side, which is fitted in front of a cardioid microphone. Sound waves arriving from the front enter the end of the tube and also all the holes. All arrive at the same time at the diaphragm to produce maximum output.

When waves arrive from the side, they enter the end of the tube and the holes as before. but there is a difference in the length of the paths travelled. Those entering the holes nearest the diaphragm travel a shorter path than those entering the end and the further holes.

There is thus a series of delays and phase differences which produce partial cancellations. As there are multiple paths through the various holes, cancellation occurs over a range of frequencies from high down to where the tube length equals half a wavelength. A high degree of side rejection thereby occurs, resulting in a polar response of a narrow forward lobe.

Below the lowest cancellation frequency the device reverts to the basic characteristic of the microphone capsule to which it is fitted, which is usually a cardioid. Some gun microphones have used an omnidirectional transducer. In this case the device reverts to an omnidirectional at low frequencies with no side rejection at all.

A 24 in. (60 cm) tube is effective above 100 Hz and gives about 10 dB rejection. We have seen that the cardioid microphone could be placed at 1.74 times the distance from the source for the same pickup of indirect sound, and a hypercardioid twice the distance. A 24 in. gun gives the same results at around 3.5 times the distance. A smaller tube of $8\frac{1}{2}$ in. (21.6 cm) is effective only down to 1 kHz and can operate at about 2.75 times the distance. Directivity coefficients are shown in Figure 27.

The 24 in. tube is too directional to be statically mounted for public address, the speaker needs only to move his head a few inches and the sound fades. It is also very unwieldy. An $8\frac{1}{2}$ in. tube is more practical, but needs to be fitted to a hypercardioid transducer with a flat response. It too is rather over-directional

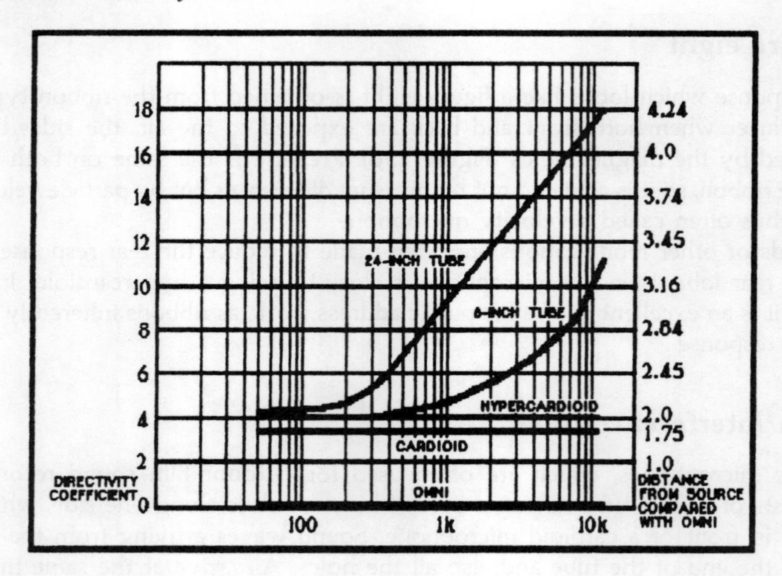

Figure 27 Directivity coefficients and distances from source compared to an omni-directional microphone.

and could result in fading as the speaker moves. An important factor is that as major feedback peaks appear below 1 kHz, these are unaffected by the smaller tube so the microphone does little to reduce feedback in spite of its greater directivity. Some short-tubed microphones using omnidirectional transducers are actually worse than a good hypercardioid for reducing feedback.

Directivity then, is an important factor in assessing its ability to reject feedback, hence its suitability for public address. The other major factor is the flatness of its frequency response.

In Figure 28 (a) and (b) the feedback level of a system is plotted on a gain/frequency chart. If the amplifier gain is turned up to that shown in (a), operation is just below feedback, and ringing is very likely. The slightest change in operation will make the system go over the top. To obtain stability and stop ringing or *incipient feedback*, the gain must be reduced to that shown in (b). It is always better to sacrifice some volume and opt for lower gain in order to avoid ringing, which degrades intelligibility and is unpleasant to listen to.

Curve (b) then, is the normal operating point for a system that has a flat frequency response throughout its range. Unfortunately this is almost never the case. Even good quality microphones have peaks in their response, some more so than others.

Together with loudspeaker and hall acoustic resonances, the frequency response is far from the flat curves depicted in Figure 28 (a) and (b). It is more like those shown in Figure 28 (c) and (d). With curve (c) the largest peak has just exceeded the feedback level. The system will therefore go into violent

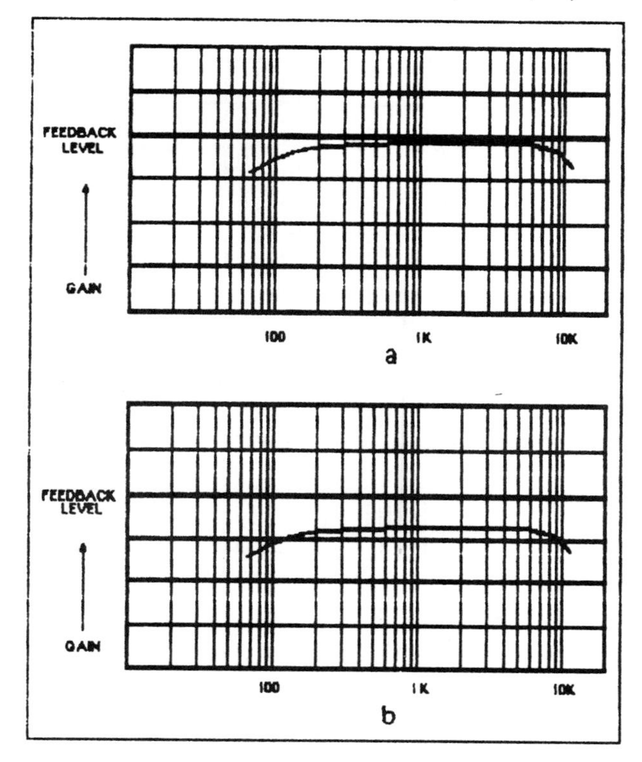

Figure 28 (a) When the gain is near feedback level ringing occurs and feedback can occur. (b) To avoid this gain must be well below feedback level.

feedback. Yet the average volume level lies well below the feedback level. To avoid feedback, the gain must be dropped to point (d), and the average gain level is thus shifted even further down.

Peaks in the response reduce the amount of gain that can be used before feedback by an amount equal to their amplitude. Furthermore, they are unstable: a sharp peak is far more likely to suddenly trigger feedback when approaching the feedback level than a gentle rise.

It follows that it is important to select a microphone that is free from large peaks in its response, and this is governed to a great extent by the type of transducer. We will now examine the main types.

Moving coil

The moving-coil type of microphone has a construction similar to that of a loudspeaker. A shallow plastic cone has a small coil wound on a former at its apex which lies within the two concentric poles of a magnet. Sound pressure waves cause the cone to vibrate and this induces corresponding voltages in the

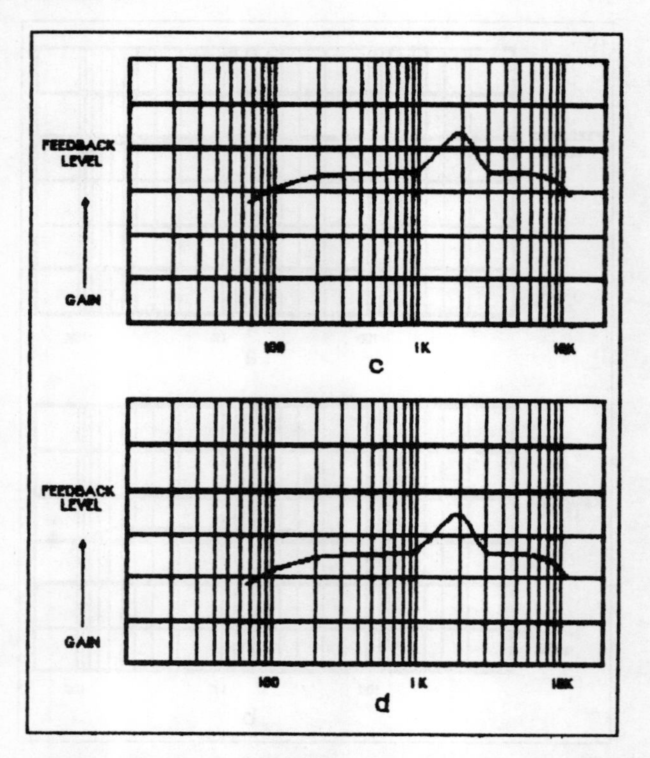

Figure 28 (c) If the tip of a peak exceeds the feedback level, it triggers feedback even though the average level is well below it. (d) To keep the tip below it, the average level must be lower still. Peaks thus set a limit on usable gain.

coil. It can be backed by a sealed chamber, making it omnidirectional, or vented to provide cardioid or hypercardioid characteristics.

The coils have an impedance of 30 ohms, which is rather low to directly drive the input of most mixers. A built-in transformer raises the impedance to 200 ohms, 600 ohms, or 50 kilohms, hence it also raises the output voltages. The 50 kilohm impedance gives a high output but suffers from loss of treble over long cables due to cable capacitance. It is only suitable for cable lengths of up to 6 feet (2 m).

The transducer is very robust and will stand a lot of hard knocks, which makes it attractive for many portable systems, especially for outdoor use. Its main snag is a resonance peak in its frequency response at around 2 kHz, which is due to the combined mass of the cone and coil.

The peak gives a 'bright' tone which is favoured by many users. It emphasizes random background noise which is usually centred around this frequency, hence the term *presence effect* used by some makers to describe it. For indoor public address use though, the peak is a definite disadvantage.

Ribbon

Ribbon microphones have a narrow corrugated aluminium ribbon supported between the poles of a powerful magnet. It is not pressure actuated because both front and back of the ribbon are exposed to the air, so the pressure is the same at both sides. Instead it is moved by the velocity of air particles as they rush back and forth past the ribbon. That is why it is often described as a *velocity microphone*. Signal voltages are generated by the ribbon moving in the magnetic field.

Impedance is very low, a fraction of an ohm, so an in-built transformer is always used to give an output impedance of 200–600 ohms. The mass of the ribbon is much lower than that of the moving coil, so it responds readily to high frequencies and fast transients. This also produces a resonance that it is out of, or toward, the upper limit of the normal frequency range. The resulting frequency response is very smooth with no large peaks, which is ideal for public address work.

Ribbon microphones are less robust than the moving-coil instrument and must be treated with respect, although they stand up well to normal usage. They tend to be expensive and few makers produce them. One such is Beyer, and their M260 ribbon which has its rear lobe suppressed to give a hypercardioid response has been the first choice for many public address systems for a number of years.

Capacitor

In the capacitor microphone a thin plastic diaphragm coated with aluminium or gold is stretched over a shallow cavity having a flat metal back plate, so forming a capacitor. Sound pressure waves move the diaphragm, thus varying its spacing to the back plate, and thereby the capacitance. A high polarizing voltage is applied to attract the diaphragm to the back plate and so keep it taut. As the capacitance changes with sound pressure, charging and discharging currents flow.

These currents produce varying voltage drops over a high-value resistor, corresponding to the diaphragm excursions, hence to the sound wave pressures. The impedance of the device is very high and if applied to a microphone cable all treble would be lost over just a few inches. So the microphone must have a built-in amplifier to serve as an impedance converter. The polarizing voltage, usually 50 V, must be supplied from the mixer and conveyed to the microphone along the cable.

This is done either by phantom powering or AB powering. With the former. the negative supply is carried by the screen of the cable and the positive by both signal conductors via the centre-tap of a pair of series resistors, or input transformer. With AB powering, the two poles are fed via the two signal conductors. A d.c. shunt path is avoided by splitting the primary of the input transformer with a series capacitor. The screen is thus no part of the circuit and so cannot inject induced hum or noise.

Some capacitor microphones have two diaphragms, one either side of the plate, facing the front and rear. The polar response of both is that of a back-to-back cardioid, producing an omnidirectional pattern. Polarizing voltage to the rear diaphragm can be switched off, so leaving the front to serve as a straight cardioid. The signal polarity of the rear section can be reversed, so giving a positive front lobe and a negative rear, thereby achieving a figure eight response, or the reverse voltage can be reduced giving a small rear lobe with some side cancellation, thus producing a hypercardioid.

So by manipulating the polarizing voltage, the whole range of polar responses can be obtained. This makes for a very versatile studio microphone, though it has little application in the public address field.

The diaphragm of the capacitor microphone is very light and free from resonances over its operating range, so giving a virtually flat frequency response, and an excellent response to high frequencies. It is the chief instrument used in recording and broadcast studios, but its high cost and the need for the polarizing voltage precludes its use for all but the most exotic public address systems.

Electret

The electret microphone is the down-market version of the professional capacitor microphone. It works on exactly the same principle, but an electrostatic charge, which is equivalent to around a 100 V polarizing voltage, is implanted permanently into the diaphragm during manufacture by heating the plates of a charged air-spaced capacitor. The need for an external voltage is thus avoided.

The charge leaks away in time, but its half-life is said to be up to 100 years. It can be much shorter though if the instrument is subject to excessive damp, such as breath condensation caused by using it too close to the mouth.

A built-in amplifier is required, but this usually takes the form of a single transistor powered by a $1\frac{1}{2}$V battery. All electret microphones thus have an internal torch cell and a switch to switch it off.

The diaphragm has to be thicker than that of the capacitor microphone in order to hold the charge, so its mass is greater and its resonant frequency lower. It thus encroaches into the operating frequency range: however, with some models it is still quite high at around 8 kHz. This is much higher than the average moving-coil unit, and it can be reduced without detriment to speech quality with a little top-cut from the tone controls.

Low cost compared to other types is a big plus, and if the model is well chosen having no peaks within the low or mid-frequencies, and it has a hypercardioid response, excellent results with low feedback is possible. The main problem with these is that the switch and battery contacts can be come noisy.

Crystal/ceramic

Certain crystals such as rochelle salt (sodium potassium tartrate), quartz and tourmaline exhibit the *piezoelectric effect,* that is they generate a voltage when physically stressed. A thin slice secured at one end with a compliant clamp and actuated at the other by a cone will thus generate an electrical signal when a sound pressure wave impinges on the cone.

This is the principle of the crystal microphone, but as these natural crystals are fragile and adversely affected by high humidity and temperature, more stable synthetic substances having the same effect have replaced them. These are barium titanate and lead zirconate.

High-frequency response is limited to around 10 kHz by the system's inertia, but can be improved at the expense of output by coupling the cone to the crystal through a reduction lever.

Two slices cemented together so that one is stretched while the other is compressed form a *bimorph.* Connected in series, these give a higher output and cancellation of mechanical strain non-linearities. Several slices are used to produce a *multimorph* or *sound cell,* in which the sound wave actuates the crystal slices directly, dispensing with the cone and its resonance.

Output impedance is very high, at around 1 megohm which gives a high signal voltage to drive valve or field-effect transistor input stages, but the low load input impedance offered by bipolar transistor circuits cause severe loss of bass. Another effect of very high impedance is excessive loss of high frequencies over cable of more than a few feet long. This is due to cable capacitance. Although once widely used for public address work, they are rarely if ever seen today. Their principal application now is for vibration and strain measurements.

Zoom microphones

Two cardioid capsules connected in opposite phase are placed one behind the other to obtain high directivity by cancellation of side-propagated sound. With frontal sounds the outputs reinforce each other when the spacing between the capsules is half a wavelength. At frequencies below this, the response falls at 6 dB/octave due to cancellation.

To avoid this loss of bass, a frequency-dependent phase shifter progressively changes the phase until the two outputs are in phase at the lowest frequency. Directivity, though, decreases to that of a cardioid at low frequencies.

The polar response is thus similar to that of the interference tube, but contained in a much shorter unit. It can be controlled down to that of the cardioid by controlling the gain of the second capsule to zero.

Output variation with the angle of incidence θ at maximum directivity is:

$$\frac{(1 + \cos\ \theta)}{2} \cos \theta$$

The instrument is mainly used for video cameras to match the operation of the optical zoom lens, but it does offer interesting possibilities for public address application, as the polar response can be tailored for a particular situation. As with other types, the microphone would need to have a flat frequency response free from major peaks.

Noise-cancelling microphones

These are usually used for announcements in areas of high ambient noise. They consist of two cardioid or hypercardioid units spaced a few inches apart and connected in anti-phase. Speech is directed at one of the units but is also picked up by the other, so producing some cancellation. As the first unit is nearer the announcer's lips than the second, there is a difference in sound pressure level, and partial cancellation occurs. Ambient sound coming from a distance arrives as a virtual plane-wave and affects both units equally. Cancellation is therefore total (Figure 29).

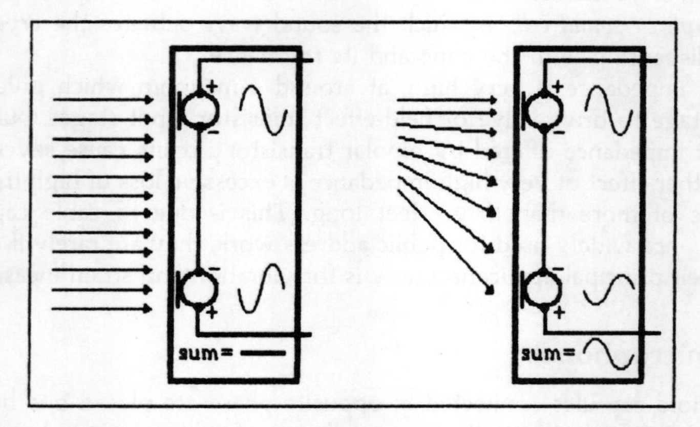

Figure 29 Noise-cancelling microphone. Two anti-phase units produce equal and opposite signals from distant sources. Nearby sources affect one more than the other so cancellation is partial.

Speech wavelengths that are shorter than the distance between the two units are not cancelled, only the longer ones. Cancellation thereby increases with wavelength, so the microphone has a falling bass response. This is no disadvantage for spot announcements in noisy conditions, as increased crispness is of greater value than high fidelity.

Tie-clip microphones

These small microphones are fitted to the tie or lapel of the performer by a clip. To achieve small size and weight they must be either capacitor or electret units. In the latter case they are usually powered by a small button cell such as used in hearing aids and watches. These have a life of between 5 000 and 10 000 hours.

Most tie-clip microphones are omnidirectional, which precludes their use for public address work; attempts to use them often end with unacceptable feedback problems. There are some models though that have a cardioid response and can be used successfully. This is especially so as the small diaphragms tend to have low mass and so have a high resonant frequency. Feedback due to this can thus be controlled with a little treble cut.

The two types of polar response can usually be distinguished by their shape. The omnidirectionals are cylindrical with end holes or grille but no other apertures, whereas the cardioids have a small ball end. The rear vent is in the base of the ball. Cardioids are thus slightly larger, but by no means obtrusive.

Boundary microphones

Boundary microphones are a recent innovation, being produced commercially in the early 1980s. They were developed and manufactured by Crown, who used the proprietary name *pressure zone microphone* (PZM). Other manufacturers are now making them under different designations.

The operating principle is acoustic, so microphone capsules of varying type and quality may be used by different makers. Caution must thus be exercised, as disappointing results may be due to an inferior capsule or its housing rather than any fault of the basic principle. Like other types there are good and bad examples.

Undoubtedly the boundary microphone offers many advantages over conventional instruments although it may not be the answer to every microphone problem. Being in its infancy, there is still much to learn as to its most effective use.

When a sound pressure wave meets a large hard smooth surface such as a wall, it is reflected from it. A microphone placed near such a surface thus receives two pressure waves, one direct, and shortly afterwards, the reflected one. At a frequency having a wavelength of half the distance from the microphone to the surface, and those that are three times, five times, seven times ... that frequency (the odd harmonics), the reflected wave is in phase with the direct one, so there is a pressure reinforcement and a peak in the response at each frequency.

At the frequency corresponding to a quarter wavelength, and twice, four times, six times ... that frequency, (the even harmonics), there is cancellation and a dip in the response. The result is a comb filter effect.

In a recording studio where movable baffles are often used to prevent leakage from one sound field to another in multi-microphone set-ups, the comb filter effects resulting from such boundaries can be a major problem. In a public address installation, wherever a microphone is used near a large reflective object such as the back wall of a platform, a speaker's rostrum or even the floor, a comb filter effect is obtained.

If the microphone is moved closer to the surface, the half-wavelength is shortened and the frequency at which the comb filter effect starts is raised. If it is moved very close, the starting frequency can be raised above the usable range of the microphone and the comb filter effect disappears.

In addition, the direct and reflected pressure waves are virtually in phase at all frequencies, so reinforcement occurs over the whole range. This doubles the overall sensitivity, giving an acoustic gain of 6 dB.

The distance from the surface at which virtual in-phase reinforcement occurs is only a small fraction of the wavelength, and it becomes smaller as the frequency rises. So to obtain the effect at highest frequencies, the distance must be minute. For the phase difference to produce a loss of only 1 dB, the distance between boundary and microphone diaphragm must be $\frac{1}{13}$ of the wavelength. At 20 kHz, this is just 0.052 in (1.32 mm). For a drop of 3 dB the distance must be $\frac{1}{8}$ of the wavelength which at 20 kHz is 0.085 in (2.16 mm). For a loss of 6 dB the distance is $\frac{1}{6}$ wavelength which for 20 kHz is 0.11 in (2.79 mm).

These losses are subtracted from the 6 dB gain, not from the original microphone sensitivity, and are only for the highest frequencies. At all lower ones the phase difference is small and so also is the loss.

Boundary microphones designed to exploit this effect have their diaphragms facing and mounted very close to a metal plate which functions as the boundary. The area of the plate is extended by mounting it on a table, wall, floor or other large convenient surface which then serves as the main or primary boundary.

With a conventional microphone, high-frequency sounds coming in off-axis, that is at an angle to the diaphragm, arrive at one side of the diaphragm before the other. There is thus a phase difference across the diaphragm resulting in a reduced excursion and lower output. Low frequencies are unaffected because of their long wavelengths, which produce insignificant phase differences. The audible effect, which is known as *off-axis coloration,* is that off-axis sounds are less sharp and clear than those normal to the diaphragm.

With a boundary microphone, the proximity of the plate, which blocks and reflects incoming pressure waves, reduces directional effects and phase differences across the diaphragm. The result is that sounds from all angles within the polar response sound the same and there is no coloration of off-axis sounds.

The polar response is hemispherical, the microphone picking up sounds from all angles on the microphone side of the boundary. The amplitude is uniform over almost the whole hemisphere, dropping about 6 dB only at the extreme edges, that is at angles from 170 to 180°. The frequency response over the hemisphere is also very consistent, with little difference between high and low frequencies, a characteristic not obtainable with any other type of microphone. The polar response can be modified to make it more directional by masking one part of the plate and unit with acoustic foam or carpeting. Alternatively, some models are designed to have a more directional response.

Pickup of sound from the opposite side of the boundary surface is dependent on frequency and the size of the boundary. At long wavelengths compared to the boundary, pressure waves from the other side are diffracted around it and are picked up by the microphone. The device is thus omnidirectional in its response to these.

Short wavelength sounds from the rear are not diffracted and so are rejected. Rear rejection frequency response thereby depends on boundary size. Given a square panel, for 3 dB rejection, the frequency is $246/d$, and the 10 dB rejection frequency is $985/d$ in which d is the dimension of one side in feet ($75/d$ and $300/d$ respectively for d in metres). Thus a 2 × 2 ft (0.6 × 0.6 m) panel rejects 10 kHz by 20 dB, 500 Hz by 10 dB, and 125 Hz by 3 dB.

The size of the boundary also determines the frequency response of frontal waves, and is to the microphone what a baffle is to a loudspeaker, though with a somewhat different effect.

Pressure waves arriving from the front are diffracted around the boundary when the wavelength is six times or more the size of one dimension of a square boundary. There is thus no reflected wave, so the 6 dB gain disappears and the sensitivity is just that of the unmounted capsule. The effect is that of a shelf filter, the gain is flat down to the critical frequency, then falls 6 dB to remain at this lower level until the bass roll-off of the capsule is reached. The frequency is equal to $188/d$, where d is one side of the square in feet ($57.3/d$ for d in metres).

One of the most remarkable features of the boundary microphone is its reach. In the near field, the output falls as distance from the source increases in a similar manner to a conventional microphone but in the free field beyond, where there is normally a 6 dB drop for a doubling of distance, the output barely changes. This, together with its hemispherical polar response, enables considerable mobility within its range.

For music, the size of the boundary should be about 4 ft × 4 ft (1.2 m × 1.2 m). For speech, a square 2 ft × 2 ft (0.6 m × 0.6 m) should be adequate. Panels this size may seem obtrusive, but clear acrylic plastic can be used, which is visually unobjectionable. If the edges of these catch the light or show up white, they can be painted black to make the panels even less conspicuous.

Diffraction around the edges of the boundary can give rise to comb filter effects, but these can be minimized by placing the microphone off-centre. A rectangular or asymmetrical shape is better still, just as it is with a loudspeaker baffle. The worst shape is a circle, as then there is an equal path to the edge at all points from the microphone.

According to the application, panels can be supported from the ceiling by piano wire, or made free-standing. Otherwise, the microphones can be installed in table tops, in lecterns, on walls, or on the floor. In the latter case, it should be mounted on a wooden panel if it is to lie on carpet, as carpet is a poor sound reflector. With a wooden or lino-tiled floor, the microphone can be just placed as it is.

For conferences a single boundary microphone will pick up around eight persons at a table if installed at the table's centre. For larger numbers more microphones may be required.

An alternative is a microphone placed in the corner between two walls and the ceiling. For this position it will be necessary to remove the plate so that the capsule can be aimed into the corner. A gain of 12 dB higher than the normal 6 dB, a total of 18 dB, is obtainable. Reverberant sound is boosted by only 9 dB because of its random phase, so there is a total of 18 − 9 = 9 dB advantage in direct to indirect sound ratio compared to an omnidirectional unit. Compared to a hypercardioid the advantage would be less, but a single hypercardioid could not be used for such a purpose. For large conference rooms, one microphone in each corner may be required.

For the theatre, three boundary microphones placed across the stage at the footlights will pick up most of the action. If the systems is stereo, the centre microphone should be replaced by a V-shaped stereo pair. This avoids anomalies for members of the audience seated on the extreme sides. For a deep stage, other boundary units may be installed toward the rear. Scenery can prove an ideal boundary mounting surface.

Theatre auditoria are usually highly absorbent, and loudspeakers out of sight of the microphones. When professional actors who are accustomed to throw their voices are on stage, boundary microphones can be very effective for reinforcement.

For a choir, a single boundary microphone, angled downward on an acrylic panel, suspended over the choir and to the front of it, will pick up all sections, even the back row, due to its long reach. Conventional microphones usually fade the further sections. For stereo, a spaced pair, single panel bipolar or V panels can be used. Floor mounting is another possibility, especially if the choir is arranged in tiers.

For public speaking, a singe boundary microphone can be fitted to the rostrum or lectern. Some models can be flush mounted to give a low profile. Notes or papers laid over the microphone do not affect its performance providing they are not too thick. Alternatively the microphone can be mounted on the floor a little to one side of the rostrum so that it is not masked by it. If the floor is carpeted, a wooden panel should be laid beneath the unit.

From the foregoing it can be seen that the boundary microphone has many advantages over existing microphone arrangements; however, it does have one major drawback, and that is feedback. The hemispherical polar response is far less directional than a hypercardioid or even a cardioid, and its long reach makes it vulnerable to loudspeaker sound from the auditorium. In recording or broadcast studios this is no problem, but for public address work it is a different matter.

In smaller halls with lively acoustics and where feedback is normally a problem, it is unlikely that standard boundary microphones would be usable. It may be possible to use a unit for rostrum speaking but larger coverage would almost certainly produce feedback at a low level because of the low directivity of a hemispherical response.

A more recent development which shows great promise is the hemicardioid model. The cardioid response is usually thought of as two-dimensional as it is drawn, but it is actually three-dimensional, being the same in the vertical as well

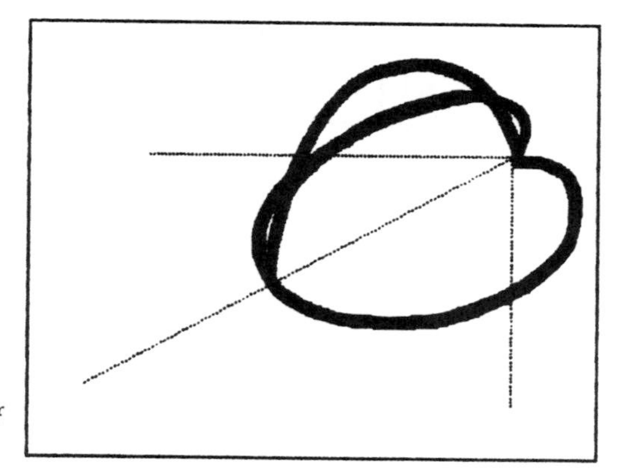

Figure 30 Three-dimensional polar diagram of hemicardioid boundary mic.

as horizontal plane. The hemicardioid is not so easy to visualize but has a full cardioid response in the horizontal plane, with half a cardioid in the vertical, somewhat like half an apple lying cut side downward (Figure 30).

A typical measured response in the horizontal plane is − 3 dB at 60°, − 6 dB at 90°, and − 14 dB at 180°. In the vertical plane there is a deviation from the strict cardioid pattern, being + 1 dB at 45° and − 3 dB at 90° which is almost level with the support surface.

The microphone is thus well suited for table discussions. Providing the participants remain within about 45° either side of the axis line which is a spread of 90°, they have freedom of head and body movement.

Being half a normal cardioid, the directivity is twice as great and so 6 dB extra gain before feedback can be obtained compared to a cardioid at the same distance. It also gives 4 dB more than a hypercardioid. So rather than causing a problem, the feedback situation is actually improved. There is though a trade-off. The hemicardioid does not appear to have the same 'reach' as the standard boundary microphone, and signal pickup falls in proportion to distance just as an ordinary microphone. It is thus not suitable for operating at a distance. For table use though it appears to be the ideal solution.

Radio microphones

The microphone itself can be any of the types and polar patterns we have described. There are two formats: the complete radio microphone in which the transmitter is inside the handle, and the pocket type of transmitter into which an ordinary microphone can be plugged. The latter used with a tie-clip microphone affords the ultimate in hands-free mobility.

Most of the cheap radio units offered for sale operate on illegal frequencies or are not approved by the Radio Communications Agency. Those approved must satisfy stringent requirements as to frequency stability and spurious emissions as laid down in specifications MPT1311 or MPT1345. This has resulted in approved equipment being rather costly in the past. However, there are now models that meet the requirements and are more moderately priced. This enables radio microphones to be used in more modest installations.

If more than one microphone is required, they should operate on different frequencies and each have its own receiver. The receivers are usually more expensive than the microphones. Each has a volume control and an audio output that can be applied directly into a power amplifier or into an auxiliary input of a mixer.

The frequencies are 173.8 MHz, 174.1 MHz, 174.5 MHz, 174.8 MHz and 175.0 MHz, using frequency modulation. These are wide-band for high-quality work and have a maximum of 2 mW r.f. output. There is also another band consisting of the frequencies 174.6 MHz, 174.675 MHz, 174.77 MHz, 174.885 MHz, and 175.02 MHz. These are for narrow-band operation with a maximum of 5 mW, for less demanding applications. Licences were at one time required for all radio microphones, but this is no longer the case with the above frequencies and powers on mainland Britain. Licences are required in the Isle of Man and the Channel Islands at the time of printing.

Higher-powered units with a maximum of 10 mW for hand-held operation or up to 50 mW for body-worn instruments operating on the same frequencies require a licence.

Having considered the different types of microphone we will now describe the various parameters and specifications used to describe them.

Microphone sensitivity

Various methods of specifying sensitivity have been used. All relate electrical output to a given sound pressure. When measured without a load in a free field, the term *free field sensitivity* or *response coefficient* is used. The unit is the mV/pascal or its equivalent the mV/(newton/m^2). This replaces the mV per μbar and the mV/(dyne/cm^2) used formerly.

Sometimes the output is expressed in minus dB. This has reference to 1 V, so -60 dB $=$ 1 mV. It should be qualified by the suffix V, thus dBV or the appended qualification (0 dB $=$ 1 V) but these are often omitted. Signal voltage varies according to the microphone impedance, so the off-load voltage is not an informative description of sensitivity unless the impedance is known. Comparison can thus only be made between models of the same impedance.

To take impedance into account, an output power rating is often used instead of voltage. The usual reference standard is the milliwatt into 600 Ω. The appended description is: (0 dB $=$ 1 mW into 600 Ω) and the term is dBm. The voltage at 1 mW is 775 mV.

The following are different ways of expressing the sensitivity of the same microphone:

0.1 mV/μbar
0.1 mV/(dyne/cm^2)
− 80 dBV/μbar (0 dB = 1 V)
− 58 dBm/(10 dyne/cm^2) (0 dB = 1 mW into 600 Ω)

1.0 mV(N/m^2)
1.0 mV/Pa
− 60 dBV/Pa (0 dB = 1 V)
− 58 dBm/Pa (0 dB = 1 mW into 600 Ω)

Impedance

Microphone impedances are normally one of the following: 30–50 Ω; 200 Ω; 600 Ω; 1000 Ω or 47 kΩ. The suitability depends on the application. One determining factor is cable capacitance. Its reactance decreases with cable length and increasing frequency, so long runs impose a low-value shunt across the output of the microphone which decreases further as the frequency rises.

With high-impedance microphones this frequency-sensitive shunt produces a falling high-frequency response. With low-impedance models, the value of the shunt is too high to have a significant effect. So to maintain the maximum high-frequency response with long cables low impedance is necessary and is commonly used in recording studios. The snag is that low impedance means low signal voltage, with the possibility of degraded signal-to-hum and signal-to-noise ratio. It also requires high-gain, low-noise input circuits.

High impedance, while offering a high signal voltage, results in noticeable loss of treble even over moderate cable lengths. The optimum for public address work is thus in the mid-band of impedances 200 − 1 kΩ.

The typical capacitance of microphone cable is of the order of 200 pF/m. Figure 31 shows the reactance of cable lengths up to 50m at six frequencies from Hz to 20 kHz. The shunting effect of a length of cable can thus be determined.

Frequency response

Extended treble or bass frequency response is unnecessary for public address work, but as noted before, it should be as flat as possible with no large peaks. Published response curves are averaged over a number of units, or may be based on the best of a batch. They should be treated with caution and regarded only as a guide. Better instruments come with an individual chart.

Pressure gradient units often show several bass curves for different source distances to allow for the proximity effect. Also shown in many plots are two treble curves giving a different on- and off-axis response. Some units have a bass-cut switch.

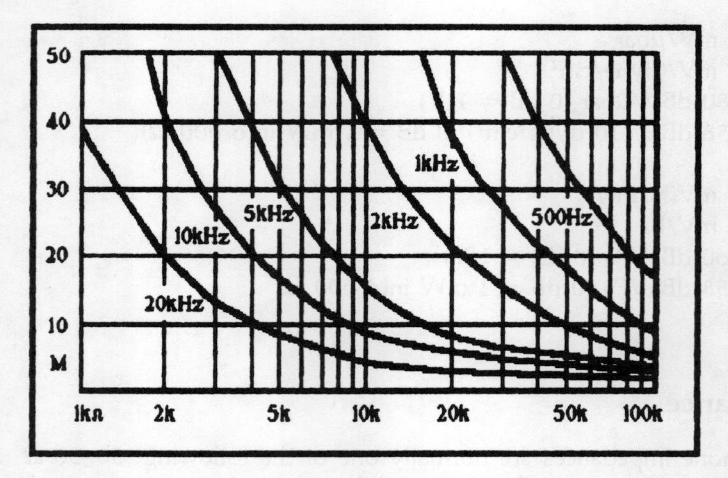

Figure 31 Reactances of 200 pF/m screened cable at various lengths and frequencies.

In all cases care must be taken to take note of the vertical calibration. Some makers used a cramped calibration which makes the response curve look a lot flatter than it really is. This is shown in Figure 32, where the same curve is shown with two different vertical calibrations, of 1 dB and 4 dB per division. Needless to say the former is rarely used, but the latter is common.

Noise

Thermal agitation in conductors and fluctuations in internal pre-amplifier and impedance-converter currents produce self-generated noise in microphones. There are several ways of expressing this. One is its measured value in microvolts. As some noise frequencies are more obtrusive than others the value can be measured after filtering according to a weighting curve such as the DIN 45 405. It is then called the *weighted noise figure*.

The voltage can be expressed as though it were generated by an external sound, thus allowing a direct comparison of levels to be made with received sounds. This is known as the *equivalent noise*, and like other sound levels is specified in decibels relative to the hearing threshold of 20 μPa. The formula for calculating it is:

$$20 \log_{10} \frac{V_n \times 10^6}{V_s \times 20}$$

where V_n is the noise and V_s is the rated sensitivity, both in microvolts.

Another method of specifying noise is the *signal-to-noise ratio*. Unlike an amplifier there is no maximum signal to use as a reference so it is fixed at 1

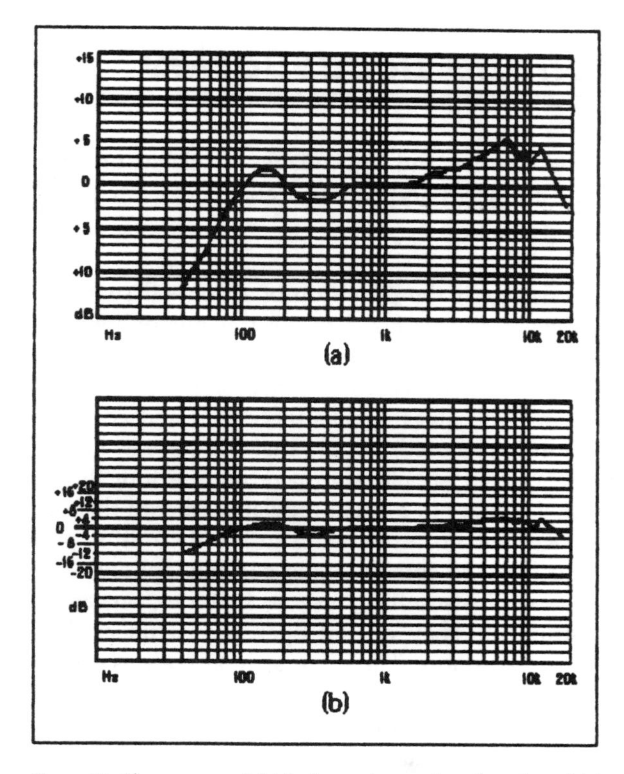

Figure 32 The response of (b) looks much smoother than that of (a). Look again at the vertical calibrations–they represent the same range. Always check the calibration carefully!

pascal, which corresponds to an SPL of 94 dB. Conversion from unweighted noise level to signal-to-noise ratio or vice versa can be simply performed by subtracting from 94. Typical weighted noise values are: capacitor 3.6 μV; moving coil 0.26 μV. The higher value for the capacitor is due to the internal amplifier.

Microphone noise is not a consideration with public address systems unless it is excessive due to a fault.

5 Microphone accessories

There are many accessories and appendages available on and for use with microphones. Some are necessary, others useful, and a few just gimmicks.

On–Off switch

The first and obvious one is the provision of an on–off switch on the body of the microphone. One is always fitted to electret microphones except tie-clips, where it serves the purpose of switching off the internal battery. Tie-clip units usually do not have a switch because of their size, but unscrewing the body, which is normally in two halves, breaks the circuit and conserves the battery, when not in use. It may be noted though, that the current drain is so small that, even if left on, the battery should still have quite a long life. Some batteries actually last longer if subject to a small regular discharge than if not used at all for long periods. Like humans, exercise, even if only a little, prolongs life.

Switches are also fitted to some moving-coil microphones. These are not to conserve battery power because they use no batteries, but are for the purpose of muting the instrument. They do not open the circuit as do most switches, but short-circuit the transducer. There are several reasons why this is done. If the microphone was open-circuited while connected to an amplifier, hum would almost certainly be picked up on the cable. By shorting out the microphone, the input circuits are also short-circuited and hum and even noise generated in the input circuit is reduced.

Switch contacts oxidize with time and can become noisy when used in low-level signal circuits. But noise can only be produced when the contacts are in circuit; if they are open there is no possibility of noise. So, an open-circuit switch when the instrument is in use can cause no problems.

Noisy switch and battery contacts can be a problem with electret microphones. Oxide can be cleaned off by applying a proprietary switch cleaner or carbon tetrachloride, then vigorously operating the switch. However, they will soon oxidize again unless protected. This can be done by a smear of grease. Alternatively, a drop of oil on the contacts will often clean and also preserve them. This may seem strange as oil is an insulator. However, when wiping contacts close the oil is pushed out of the way leaving bare metal, but when they open the oil spreads back to cover the surfaces and so protects them from atmospheric moisture and pollutants.

Another reason for short-circuiting the transducer in the 'off' position is that it protects it in transit. This may not be important for the robust moving-coil unit, but could be in the case of the more fragile ribbon. Any excursion of the ribbon generates an e.m.f., which produces a current through an external circuit. The current generates its own magnetic field which opposes the original motion, so acting as a brake.

If there is no external circuit there can be no current, and so no damping. The short-circuit provided by the switch in its 'off' position thus serves as a very low resistance shunt circuit, and thereby affords maximum protection against shock.

However, the on–off switch is not an unmixed blessing, it can be off when it is supposed to be on. It is one more thing to have to check before going 'on the air'. Also when using the instrument as a hand microphone, nervous performers are apt to finger its contours. If that contour happens to be the switch, they are likely to switch themselves off without knowing it. The operator may not be aware of what has happened, and nearly has a heart attack as the system goes dead!

Speech–music switch

Another switch found on some microphones is one that selects either 'speech' or 'music' operation. This is really a bass cut switch, the cut being in the 'speech' position. Bass cut can be beneficial, especially when the microphone is being used close to the lips, as this mode of operation emphasizes bass. Even when used at a normal distance, hall and loudspeaker resonances, which ruin intelligibility and encourage feedback, can be tamed to some extent with a little bass cut, although it should not be overdone.

However, here again there is the possibility of the switch being operated inadvertently by a performer, furthermore bass cut may be needed for one speaker but not for the next. Really, all control of tone should be fully in the hands of the public address operator, and adjustable by him if needed during the programme. Any bass cut should thus be effected by means of mixer or amplifier tone controls.

Both on–off and speech–music switches are generally more of a nuisance than they are worth, and with the exception of electret microphones, should be avoided in instruments used for public address work. Even with electrets it is sometimes possible to modify them to cut out the switch and complete the battery circuit by a link in the microphone connector. Alternatively they may be A B powered from the mixer.

Impedance switch

Yet another form of switching sometimes encountered on microphones is that which enables two different impedances to be selected, usually high (47 kΩ) and 1 kΩ. This is achieved by an internal transformer to provide the high impedance. It is selected by either a switch, or more usually, by a connector that can be fitted in more than one position. In this case a marker lines up with one out of two spots on the microphone casing.

High impedance is never used for public address work for reasons previously given. Dual-impedance models are therefore unnecessary and can cause confusion, with the possibility that the wrong impedance will sometime be selected. Unfortunately, some excellent models that are eminently suitable for their good feedback characteristics are available only with dual impedance.

Other refinements

Capacitor microphones usually have a high output because of the internal preamplifier, and so could overload mixer input stages. For this reason some have a built-in attenuator to reduce the output as required.

Another feature found with some professional models is a switch that can change the polar response from omnidirectional, or cardioid, or hypercardioid to figure eight. This is done by varying the polarizing voltage to one of a pair of diaphragms so altering its response.

Many better microphones have shock-absorbing mounting of the transducer within the casing to reduce handling noise, and some have built-in pop filters, although these are of limited effectiveness.

Connectors

All professional microphones have detachable leads with connectors to connect them to the microphone. Only the cheaper models have leads that are permanently wired in. These are usually short, about 6 ft (2 m) in length, and are single-conductor-with-screen, so forming an unbalanced circuit. For longer runs an extension with connectors is needed which also must be unbalanced. Breaks in the cable within a few inches of the microphone, where they usually occur, cannot be quickly repaired with wired-in cables. With detachable leads, a replacement can soon be fitted. The only type of microphone that must be used with a wired-in lead is the tie-clip, as there is no room for a bulky connector.

There are various types of connector used for the lead-to-microphone connection. The most common are the XLR, DIN, Tuchel, and a nameless though popular audio connector. These mostly have three pins, so affording balanced circuit operation, but the latter audio type is available in three- or four-pin versions (it can also be obtained in up to eight-pin configurations, but these are not used for microphones). The four-pin reversible version is often used for impedance changing.

The standard connections are in Figure 33. Connectors used for the amplifier end of the cable are normally either the jack plug or the XLR. The jack has either two or three contacts for unbalanced or balanced operation. The XLR is the same as that used for the microphone end of the cable; both ends are of the female type, while the sockets on the microphone and the amplifier are male.

The jack has been used for many years and has proved reliable in the past, but recently they seem to be prone to a number of faults. Some having shiny plated barrels look impressive but have poor electrical contact. Others have been found with a loose or high-resistance connection between the tip-contact and

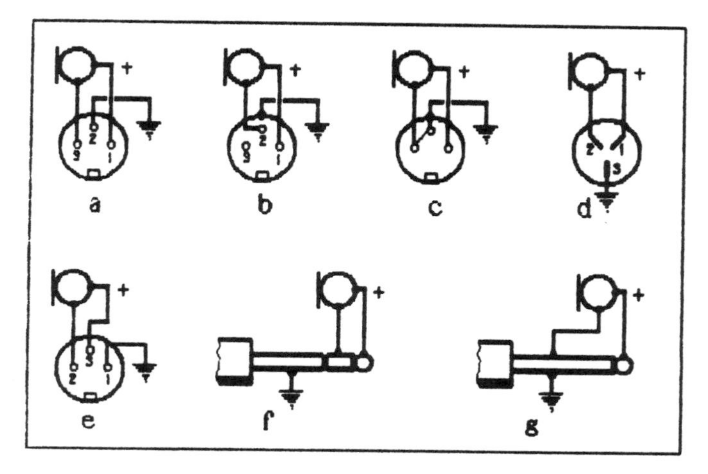

Figure 33 (a) DIN balanced; (b) DIN unbalanced; (c) alternative DIN unbalanced; (d) Tuchel balanced; (e) XLR balanced; (f) Jack balanced; (g) Jack unbalanced. Viewed from free end of microphone pins and solder tag end of plugs.

its riveted terminal. All told, the XLR plug is rugged and is the most reliable; many having gold-plated pins for low-resistance noise-free contact.

Microphone stands

There is a variety of microphone stands available (Figure 34). The first and most used is the floor-stand, consisting of a telescopic set of tubes, which are usually chromium-plated. The majority have two sections which give a maximum height of around 5 ft (1.5 m) and a minimum height of just over 3 ft, (1 m). There are also three-section models that extend to about the same height, but reduce down to about 2 ft (0.6 m).

Each section is clamped in place by a twist ring on the section below it. In some models the sections are pneumatically supported, when raised, by air trapped in the section below it. Lowering is accomplished by firmly pushing the section down, thereby forcing air through a release valve.

Stand bases can be circular or shaped as three pods or limbs, or they can have three short legs set almost horizontally. Some have detachable legs or ones that can be folded back against the stand for transportation. Solid bases should be heavy, or the legs long enough to give a high degree of stability.

As the stand and its fittings are prominently on view they should be visually attractive with a 'quality' appearance. The audience will judge the standard of the whole system not only by how it sounds by on its appearance. A tarnished or nondescript stand will give a poor image even though the sound quality may be excellent.

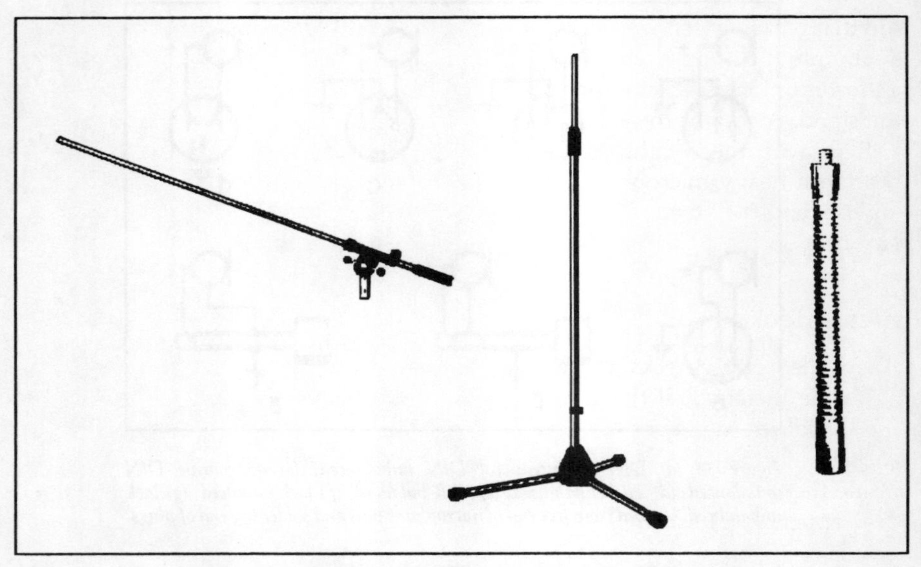

Figure 34 Microphone boom arm, stand and swan-neck.

Table stands are smaller versions of the floor stand. Some have telescopic short sections, but others are fixed in height. Many are just a base with a short stem in either metal or plastic and there is quite a variety of shapes and sizes.

Stand threads

Over the years a number of different thread sizes have been used for connecting the microphone or clamp to the stand. This lack of a standard may pose problems when the chosen instrument has a different thread to that of the stand.

The most common sizes are $\frac{5}{8}$in (1.6 mm), $\frac{1}{2}$in (13 mm), and $\frac{3}{8}$in (9.5 mm) with 27, 26, and 23 threads per inch (10.6, 10.2, 9.1 per cm). Less common but also used are $\frac{5}{16}$in (8 mm) and $\frac{3}{4}$in (19 mm). As a rough guide, the $\frac{5}{8}$in size is mostly found on American and Japanese equipment, the $\frac{1}{2}$in on British, while the $\frac{3}{8}$in appears on European instruments and stands.

A range of thread adaptors having a male thread of one size and a female of another is available to enable a match of most sizes to be made.

Boom arms

It is usually necessary for the stand to be placed a short distance from the speaker to allow for a desk or rostrum. The microphone must thus be brought nearer, and one way of doing this is by means of a boom arm. This is a metal rod that is mounted on the top of the stand and that can be extended horizontally at any required angle. Its clamping point on the stand is also adjustable to give a shorter or longer reach with a corresponding longer or shorter overhang behind the stand. A weight to serve as a counterbalance is often fitted to the overhang.

Even with the weight, stability is poor when the arm is extended fully and bearing a heavy microphone such as a moving-coil unit. Full extension is thus unwise, and the stand should be orientated so that one leg lies underneath the arm or as near so as possible; this will give the greatest forward stability .

Swan-necks

Also called goose-necks, these are an alternative to the boom. They consist of a flexible metal spiral that can be bent to any position. They are mounted on the top of the stand and a microphone clamp fitted to the free end. Some have sockets into which the microphone is plugged, in which case the cable runs inside the swan-neck and exits through a hole near its end. These present a very neat appearance but the microphone cannot be used when detached from the stand.

Swan-necks range from 9 to 24 in (288 to 610 mm) in length, but the longer ones are not suitable for carrying heavy microphones as they tend to wilt under the weight. They also develop rather unsightly bends along their length from previous positionings, and need to be straightened out from time to time. A 15 in (38 mm) is the maximum for a moving-coil or ribbon microphone. A 9 or 12 in (228 or 305 mm) unit is usually adequate and is the best for most applications.

Adjustment for angle and height is quickly effected without unscrewing clamping rings, and the absence of any rear extension can be an advantage when operating in a confined space. So a swan-neck is to be preferred for short-reach applications, but those for which a longer reach is required are better served by a boom arm.

One point to watch with swan-necks is that they should never be bent at a sharp angle as this can pull the spirals apart; they should always be set in a gentle curve. Another point is that some are much stronger than others, and weak ones, even if short, will not bear the weight of a moving-coil or ribbon microphone.

Microphone clamps

The simplest version of the microphone clamp is a cut-away plastic tube or ring which is of slightly smaller diameter than the microphone so that it grips it when it is pushed in. As microphone cases differ in diameter, the clamp must be of the correct size for the microphone in use. In addition to cylindrical models, some microphone cases are tapered and others are square-sectioned. These must have the appropriate clamp.

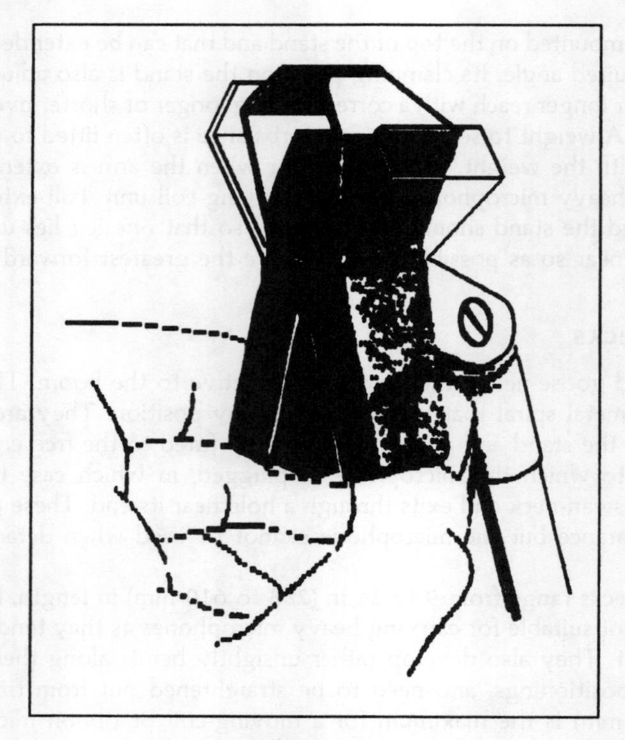

Figure 35 Adjustable microphone clamp.

A more versatile clamp is the one that has two spring-loaded clamp pieces that are opened by squeezing the lower tabs (Figure 35). These will accept most microphones and also make them easier to detach and replace. The only disadvantage is that they are less elegant in appearance than the conventional fixed clamps.

Wind shields

These are described as pop shields because they minimize the pops or explosive consonants P and B when used close to the lips. The simplest type is a foam plastic hood which fits over the end of the microphone. When used in the open air, wind noise, which gives the sound of irregular roaring, is reduced by shielding the diaphragm from direct air-particle velocity.

Large shields give a wind-noise reduction of some 20 dB, but the smaller slimmer ones often used give much less. If a near total reduction is required, the shield must be more elaborate. These consist of large closely woven wire-mesh cages with internal layers of silk and other materials. They are designed to have only the minimum effect on the frequency and polar response of the instrument.

Microphone care

Microphones are delicate instruments and should be treated as such, although many will stand up to a surprising number of hard knocks. When transported they should always be carried in suitable containers. The makers of the better ones usually supply a foam-lined case with each unit, although these are often made of soft plastic which soon disintegrates. For regular transportation a more durable container may have to be constructed.

When connecting to any external circuit, care must be taken that no voltage exists that could pass a current through the microphone. A small current from an ohmeter for testing is permissible, but nothing larger. A low resistance will be indicated in the case of a moving-coil instrument or one that has a matching transformer, but not in the case of a single-impedance electret. A low reading with one of these would indicate that the series coupling capacitor was leaking.

Dampness should be avoided with all types. Rust and corrosion can result with magnetic types, and leakage of the in-built charge with electrets. Dust has little effect on a diaphragm or cone but can cause problems with a ribbon by filling the narrow gap between the ribbon and the magnet pole pieces. Especially to be avoided are iron particles which are almost impossible to remove from the pole face without damaging the ribbon.

Batteries used in electrets need replacement occasionally although they usually last a considerable time. As mentioned before, battery and switch contacts should be treated with a spot of oil or light grease to protect them against damp and oxidization.

6 The programme mixer

It is at the programme mixer that all the inputs from the various signal sources are brought, selected, adjusted for level, tone, and special effects, combined, and ultimately fed to the power amplifiers. It is usually the most complex item of equipment in the system. Before considering the features and specifications of typical mixers we will look at the basic principles involved.

Noise

One of the major considerations is that of reducing noise to the lowest possible level. All amplifying devices presently known produce a random assortment of spurious frequencies. Because each octave has double the frequency span of the one below it, most of the noise energy is concentrated in the upper octaves and the audible effect is that of a high-pitched hiss. It is often called *white noise* because like white light, it is a made up of a large range of individual frequencies. This distinguishes it from what is termed *pink noise,* which is white noise filtered to give the same amount of energy in each octave. It thus has a larger ratio of lower frequencies, just as pink light has more long-wavelength red than white light. Pink noise is often used for various audio tests.

In the case of semiconductors, the noise arises from two main causes, the base–emitter internal resistance which generates thermal noise, and emitter current. Base–emitter resistance decreases as the emitter current increases, so noise from one cause diminishes as the other worsens. There is, however, an optimum current for each type of transistor at which noise is at a minimum. Good designs operate the transistors at their optimum noise currents, especially in input stages.

A signal-to-noise ratio is usually specified for each mixer input as a minus decibel value. This is the noise level compared to the maximum rated output. Sometimes it is specified in millivolts. The minus decibel value can be calculated in such cases from:

$$20 \log_{10} \frac{V_o}{V_n}$$

in which V_o is the maximum rated output voltage and V_n is the quoted noise voltage.

Stage gain and noise

Mixers and amplifiers consist of a chain of stages, each amplifying or modifying the signal before passing it on to the next. So the signal grows larger as it

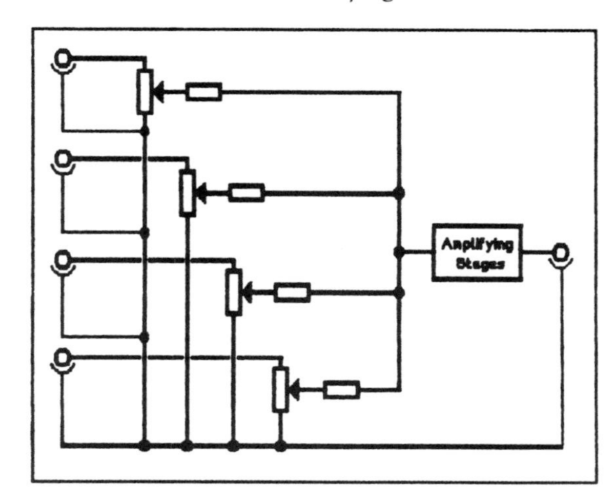

Figure 36 Simple mixer with no pre-amplific stages.

progresses, and each successive stage must be able to handle the increased level. Thus the input stage is designed to amplify very small signals, while the output stage must handle quite large ones. Those in between are designed for intermediate values.

The emitter current for each stage, which is set by the resistor values in the bias network, is chosen on the basis of the signal it must handle. The larger the signal, the higher the current. This in turn determines the gain of each stage and also the noise it generates. High currents means high noise and low gain.

As a result, the gain of each successive stage is less than its predecessor while its noise is greater. In spite of this it is the input stage that contributes most noise. The reason is because its noise is amplified by all the stages that follow, whereas the later stages, though individually generating more noise, have but few following stages to amplify it. Even those have low gain.

The input stage of each channel must thus be carefully designed using low-noise semiconductors with the minimum emitter current required to handle the maximum input signal for which it is designed.

Faders and noise

Each mixer channel has a fader to individually regulate its signal level. The output from these are combined and fed to the main amplifying chain as shown in the circuit of a simple mixer in Figure 36. It will be seen that a resistor is included in the wiper connection of each fader. These are known as *swamp resistors;* without them the output from all the other faders would be shorted out when any one is turned down. Even so they shunt the signal when grounded and so must be of fairly high value.

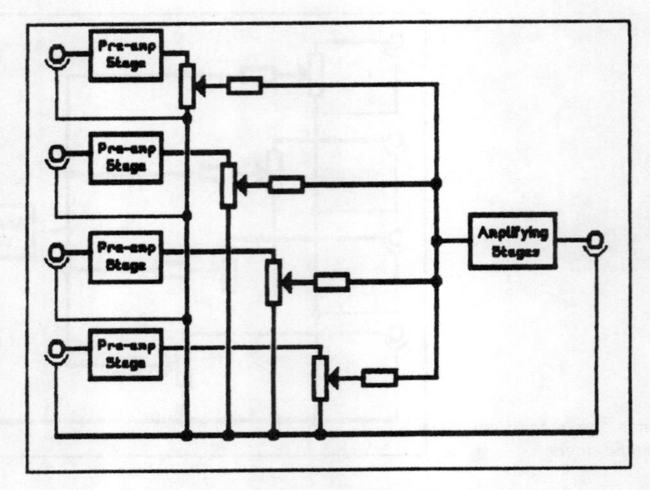

Figure 37 Mixer with pre-amplifier stages.

Being in series with the signal to the next stage, they introduce a signal loss, which increases with their value, which must thus be a compromise. Combined with the loss in the fader the result is a much reduced microphone signal reaching the first stage. The signal-to-noise ratio is thereby considerably impaired.

For this reason this simple circuit is never used. Instead, the input is applied first to an individual pre-amplifier stage as shown in Figure 37. The signal is thus amplified before it is subject to losses in the fader network, and the input stage noise is turned down with the input signal by the fader. Thus the signal-to-noise ratio is not greatly changed at low fader settings, as it is without a pre-amplifier stage.

An alternative fader arrangement is shown in Figure 38. Here the wipers of the faders are connected to the previous pre-amplifier stages and the top ends connected through their swamp resistors to the next stage. Much lower values for these can be used and the loss is much less.

A disadvantage with this circuit is that it is frequency selective, there is a bass loss when the fader is at a low setting. However, this only occurs when the resistance of the fader below the wiper position is lower than the reactance of the coupling capacitor. If the total fader resistance is not less than 10 kΩ and the capacitor value not less than 1μF, the effect will occur at only the bottom tenth of the fader travel which is well removed from normal operating positions. This arrangement has been successfully used with custom-built programme mixers.

Another method which avoids these problems is to have individual buffer stages after each fader. There is then no need for a swamp resistor with its signal loss. An alternative which is often used is to have a two-stage pre-amplifier

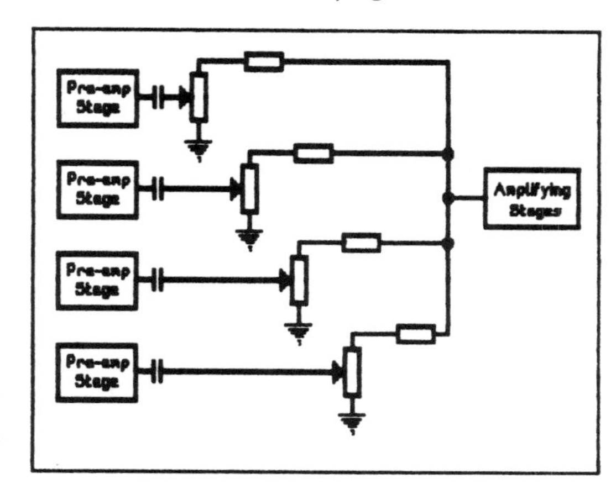

Figure 38 Fader arrangement reduces losses by permitting low-value isolating resistors. There is a slight bass loss at low fader settings.

before the fader. Swamp resistors are needed if the faders are connected together to the common amplifying stage, but a high signal level is applied, and the noise of both pre-amplifier stages is turned down by the fader. The two-stage pre-amplifier is usually directly coupled to reduce the component number.

Noise specifications

Care must be exercised when checking published noise figures, that the measuring conditions are understood. These are not always clarified by the makers, sometimes purposely, in order to make the specifications seem better on paper than they really are.

Residual noise is often quoted, and is the noise obtained with all controls including the master turned down. This is usually just the noise of the output stage and is very low, so looks very impressive. A typical value is − 95 dB. It does not indicate the level experienced in practice as the mixer is hardly much use with its master fader fully down!

Another value is that given with the master control at the 'normal' position and all input faders at minimum. It indicates part of the noise in the post-fader amplifying chain minus the input circuits. Around − 70 dB is an average value. This is the noise level obtained with no signal, such as during a pause in the programme with all faders down, and it is most noticeable because there is no signal to mask it.

A further condition often specified is with one channel at maximum and a resistor connected to its input to simulate the microphone impedance, all the others being at zero. The level experienced under these conditions, or a slightly higher level, is that to be expected during normal working. A value of − 60 dB is typical, but − 50 dB should be considered the minimum. A resistor of lower value than the microphone impedance will produce a lower noise reading, so some makers use values as low as a quarter of the specified input impedance.

Overload characteristics

The input stage is designed to handle only a small signal, because then it can have a high gain and generate the minimum noise. This is important because, as we have seen, there is a lot of amplification following it. However, this means that if a large signal is presented to it, it will be overloaded and distortion will result.

The reason for this is shown in Figure 39. In (a), the transfer characteristic of the amplifying device is shown with an input signal applied to its linear part. The output is of the same waveform, but amplified. In (b), a larger input signal is shown, which encroaches onto the curved parts of the characteristic. The resulting output has the tops of the waves clipped off, and the effect is thus termed *clipping*. Whatever the amplitude of the input signal beyond this level, the output cannot follow it but is always clipped at these same points.

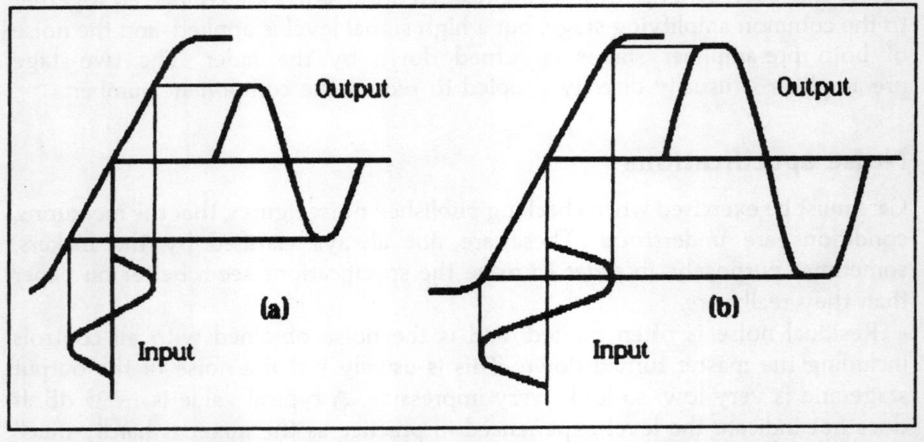

Figure 39 (a) Amplifying stage with input and output waveform. (b) Excessive input encroaches on curved part of the characteristic so producing an output with clipped peaks.

The clipped wave tops resemble a square wave. As shown in Chapter 3, a square wave consists of a fundamental frequency with a high order of odd harmonics. Clipping thus generates a large amount of spurious harmonics which impart a hard unpleasant quality to the sound.

In a practical situation, mixer microphone inputs are designed for low signal levels, which for a low-impedance microphone may be 0.25 mV. If a signal from a tape player having an output of some 100 mV is applied, there will certainly be gross overloading. Even if the tape output control is kept low it would be virtually impossible to get it low enough to avoid exceeding the input limit.

However, all well-designed inputs have an overloaded tolerance which enables them to handle signals in excess of those specified. So, a 0.25 mV input may well handle up to 5 mV or even more without overloading, although there would be an increase in harmonic distortion. The tolerance is necessary because audio signals can be unpredictable and large peaks many times greater than the average can appear without warning. Allowance must therefore be made for this in the design.

Looking at the opposite situation, if a microphone is connected to a tape input there is insufficient gain, but if usable gain is obtained by turning everything up, there is a high noise level.

These are extreme examples, but a similar situation can arise by using microphones of incorrect impedance. The problem is not so much impedance matching because maximum power transfer is not an issue. Input loads are usually many times higher than the microphone impedance to avoid signal loss through the potential-divider effect (see Chapter 3). The significant factor is the signal voltage. High-impedance sources generate high signal voltages while low-impedance ones produce low voltages.

If then a low-impedance microphone is used with a high-impedance input, the gain will be insufficient, the fader will have to be turned well up and a high noise level will result. A high-impedance microphone may overload a low-impedance input, but some signal loss due to the low load partly compensates.

The difference between 200 Ω and 600 Ω microphones and inputs is very slight and has no noticeable effect for public-address work. They can therefore be used interchangeably, a 200 Ω microphone into a 600 Ω socket and vice versa. These are the preferred impedances.

Sensitivity

The signal voltage quoted for each input denotes the voltage that will produce the maximum rated output from the mixer with the appropriate fader full up. If exceeded, the input circuit may not be overloaded due to its overload tolerance, but a later one may be. This is because the maximum rated output means just what it says; any further increase in output comes at the cost of high distortion due to overloading of a subsequent stage.

However, this can be prevented by turning the fader down, so the capability for handling high input signals without overloading remains entirely that of the input circuit. Overloading elsewhere is possible only if the fader is up too high. Some indication of whether this is happening, or is on brink of happening, can be obtained from the output meter. If it indicates a reading exceeding or close to that of the maximum specified output on peaks, the gain should be reduced. We will take a closer look at the output meter later in the chapter.

For microphones in the medium impedance range 200–600 Ω, which is the recommended range for public-address work, the sensitivity should be between 0.25–0.5 mV. Dynamic microphones, that is moving-coil or ribbons, have average outputs around 0.25 mV, while electrets are nearer 0.5 mV. As the output varies over a wide range according to the manner of use, distance from, and intensity of the sound source, these figures are very approximate. The overload level should be at least 10 mV, which is not often quoted in manufacturers' specifications. Professional mixer inputs exceed this figure considerably.

Sometimes the sensitivity is given as a minus dBV value. This is relative to 1 V, so − 66 dBV corresponds to a sensitivity voltage of 0.5 mV and − 72 dBV to 0.25 mV. Occasionally a range may be quoted, such as − 60 to − 6 dBV. This indicates the sensitivity and the overload level which in this case is 1 mV and 0.5 V respectively.

Balanced and unbalanced inputs

The microphone input can be either balanced or unbalanced. With the unbalanced circuit a single conductor is used with a braided screen. The screen is earthed, which as well as screening the 'live' conductor from magnetic hum fields, serves as the return leg of the circuit.

This is quite satisfactory for small installations and short cable runs, but is less so for larger ones. The screening does not totally eliminate stray magnetic fields, it only reduces them. So the inner conductor can still have small hum voltages induced into it.

Furthermore, hum and noise can be induced into the screen itself, which is in series with the microphone circuit. These are thus injected in series with the microphone signal into the input.

With the balanced circuit two wires are used within the screen, one the forward and the other the return. These are twisted together so that any hum field penetrating the screen affects both equally. As they are connected at opposite ends of the microphone, such induced voltages cancel. The screen is connected to earth at the mixer, and to the metal casing at the microphone, but it does not form part of the input circuit. Any voltages induced into it are therefore not injected into the input.

Another practical advantage of using twin balanced cable is that some types have internal fibre padding to give a smooth circular contour. This protects the cable from damage, and also relieves bending strain. This strain usually occurs at one point, which is within an inch or so of the microphone, where it produces eventual breakage of the conductor. Padded cables are far less vulnerable to strain because bends are not so sharp but distributed, and so they rarely break.

It is possible to use twin padded cables with an unbalanced circuit in a quasi-balanced configuration. This gives most of the advantages of balanced operation. It is done be simply using the twin conductors for the microphone circuit, with the screen connected to the earthy side at the mixer end but not at the microphone. Here it is either connected to the casing of a three-pin connector is used, or left isolated if the connector is a two-pin type.

All professional mixers have balanced inputs, these having either input transformers or integrated circuits using inverting and non-inverting inputs. If phantom or A–B powering is available for capacitor microphones, a transformer is used, possibly centre-tapped for phantom, or with split halves for A–B.

It is virtually impossible to shield an input transformer from hum fields in mains-driven equipment with ordinary metal shielding. It must be of mu-metal. Even with this, provision is often made to orientate the transformer for minimum hum pickup by means of single point fixing. However, mu-metal is a very effective screen and orientation is rarely necessary unless particularly strong a.c. fields are present.

Line inputs

Provision is made for feeding a high-level signal from a tape deck or CD player. The impedance is usually from 47 to 100 kΩ, but can be as low as 20 kΩ. Sensitivity is generally around 100 mV. Because it is high level and the output from the deck is controllable, this input is far less critical than that of a microphone.

On some mixers the input is described as a phono input, but this is misleading because it cannot be used directly from a record-player pickup. If a magnetic cartridge is used, equalization is required and the output is far too small as these deliver around 1–3 mV. If a ceramic pickup is used, the load is too low as ceramics need a load of around 1 MΩ. A CD player can be connected as these have an internal amplifier to deliver a suitable line output. However, the phono designation was around long before the advent of CD, and has always been a source of confusion. As vinyl discs are now dying out as a source of public address music the problem is likewise disappearing.

A feature with some mixers is that the inputs can be switched individually for either microphone or line. They can thus be tailored to a particular system. One may need two or even more line inputs with only one or two microphones, while another may need many microphones and no line inputs.

A further variation involves the use of slide-in circuit cards. The mixer can accommodate a given number of cards which can be microphone or line: time-signal, chime, or alarm generator. These are especially useful for large factory installations.

Tone controls

Tone controls can be very useful for public address work and are included in most of the more expensive mixers. Some of the professional models have a set for each channel, although this is rarely necessary if matched microphones are used. Microphones of different types have a different sound, and individual controls can help to give a more balanced effect. However, unless there is some unusual application it is better (and cheaper) to stick to one microphone model and have fewer tone controls.

Controls usually are treble cut and boost, and bass cut and boost. In the erstwhile days of peaky moving-coil microphones and tinny horn loudspeakers, treble cut was a big advantage. This is not so now, and bass cut is the main need. A little treble boost is actually sometimes required to improve clarity. Some mixers have only a switched bass cut.

Many mixers include a five-band equalizer. These can assist in obtaining a balanced sound and reduce acoustic deficiencies of the hall, but are of little use for reducing acoustic feedback as we shall see in a later chapter. Anything less than a 30-band third-octave equalizer is unlikely to be of much help for this purpose.

Reverberation

A feature found on some mixers and in much demand for pop concerts is *reverberation*. This adds body to vocals and makes mediocre vocalists sound much better than they are. It could be called the bathroom effect—the bath has long been recognized as the best place for singing, not because it is out of earshot of the rest of the family, but because of the reverberation produced by large areas of ceramic tiles.

Artificial reverberation can be obtained by several means. With all, the signal is delayed by a small amount, fed back to the input of the delaying device to be further delayed, then re-applied successively for many cycles. Each time the signal is attenuated it decays and eventually dies away. At the output of the delaying device there thus appears a whole series of delayed repetitions of the original, like multiple echoes. These are mixed with the original to give the effect of reverberation following the main signal. The level of reverberation can be varied, so enabling the ratio of direct and delayed signal to be controlled. Also the time it takes to decay can be controlled. A wide range of effects is thus possible.

There are three ways whereby the signal can be delayed. The original method is mechanical. A coiled steel spring has a transducer at one end to generate vibrations, and another at the other to pick them up. The vibrations run back and forth along the spring giving both delay and feedback. The timing of all the reflections is a function of the spring length, so the result does not have the random nature of acoustic reverberation with its many different timings. To improve on this, some units have three springs of different lengths operating in parallel.

The spring or springs are suspended from supports under tension. A reasonable length is required to obtain a good delay, so the unit tends to be rather bulky.

Another problem is that the device is prone to mechanical shock—any knocks when it is in operation sound like a load of bricks being dropped in a cathedral! This necessitates the unit being well cushioned with flexible supports, which add to its size.

Spring reverberation units are not much used now because of these snags and the development of electronic means of achieving the same result. One of these is the digital delay line.

With this the signal is first compressed to reduce the dynamic range to one that can be quantized by a restricted number of digital bits and after pre-emphasizing the treble it is sampled at around 40 kHz. It is then converted to binary digital words and passed along a memory store. At the other end the binary data is applied to a digital–analogue converter, passed through a low-pass filter to remove the quantization steps, de-emphasized, and the dynamic range expanded to that of the original. The output from the memory can be taken from earlier points, so giving a smaller delay as required. These are costly devices more likely to be found in recording studios than public address set-ups, for which the type generally used is the analogue delay line.

Analogue delay lines are built around what is termed the *bucket-brigade* chip, which consists of a chain of from 512 to 3328 capacitors contained in a single IC. Several can be cascaded to form longer chains if required. The signal is sampled at three times the highest signal frequency and each sample charges the end capacitor of the chain. On receipt of a pulse from a clock generator, the capacitor passes on the charge to the next capacitor, and a further pulse transfers it to the third one and so on down the chain. It emerges at the end with a delay dependent on the number of stages it has passed through and the frequency of the clock.

A problem is that although the samples come in a continual succession, a capacitor cannot receive another sample while it is transferring the previous one to the next capacitor. This is overcome by actuating alternate pairs. So capacitor 1 transfers to capacitor 2 while capacitor 3 transfers to capacitor 4. On the next pulse, capacitor 2 transfers to capacitor 3 while capacitor 4 transfers to capacitor 5. This means that only half the capacitors are conveying information at any time, and the effective length is halved. Alternate positive and negative pulses are required to trigger the odd- and even-numbered capacitors.

A further problem is it takes a long time to completely discharge a capacitor because it discharges exponentially with time. So there is always a varying residue of charge left in each as the charge passes along the line. A distorted signal thus appears at the output.

To overcome this, instead of the actual charge being passed from the first to the second, the second is fully charged initially, and then is discharged into the first. This already contains a sample, so when it is full, the second has some charge left over which is equal to that of the original charge in the first. Thus an effective transfer of the sample has been made from the first capacitor to the second without emptying either. This process continues down the line.

If this seems somewhat obscure, the bucket analogy can help to understand it. Imagine two buckets of the same size, and it is required to pour the contents of the first which is exactly one pint, into the second. It could be just tipped in, but this would leave some water in the first unless it was left draining upside down over the second for a long while. Alternatively, the second could be filled, then its contents poured into the first which contains the pint, until it is full. This should leave exactly one pint left over in the second bucket.

The signal appearing at the end of the line is a series of alternate pulses and gaps which would require a very steep, complex and expensive low-pass filter to smooth out. Instead, an output is taken also from the penultimate stage and combined with the final one. The pulses of the one occupy the gaps of the other, giving a much smoother result which needs only a modest low-pass filter.

The delay obtainable is given by

$$d = \frac{n}{2f_c}$$

in which d is the delay in ms, n is the number of stages, f_c is the clock frequency in kHz.

The clock frequency must be at least three times the highest audio frequency so the required frequency response sets the limit on the delay available. For public address 6 kHz is a reasonable upper limit, so a frequency of 20 kHz may be considered a suitable minimum. However, many devices have 'hairpin' distortion/clock frequency curves. Distortion can be as high as 10% at 20 kHz and fall to 0.4% at 40 kHz, rising again rapidly above 60 kHz. For such devices the clock frequency should be kept within 30–80 kHz to avoid high distortion, preferably within 40–65 kHz.

Delays can be varied by means of the clock frequency within these limits, but some devices have a series of taps enabling smaller delays to be selected. Otherwise the device can be chosen to give the approximate delay required, and the exact amount obtained by choice of frequency.

Devices can be cascaded to create longer delays although this increases noise. Better results can be obtained by parallel multiplex operation. This has two devices in parallel with their outputs summed, and the clock phase inverted for the second device. The input signal is thus alternately sampled by each, which doubles the sampling rate. The clock frequency can thus be halved to achieve the same signal frequency response, and thereby double the delay time without increasing noise.

Another arrangement is the differential system which has two devices in parallel as before, but with the signal in opposite phase in each. The outputs are applied to a differential amplifier. This cancels second harmonic distortion and clock glitches. Combining the multiplex and differential arrangements gives a high-quality system but uses twice the number of devices as the cascade or multiplex system by itself.

We have considered the bucket-brigade delay line in some detail because it is used in public address systems for purposes other than generating reverberation. We will discuss these later.

Cross-fading and priority

Cross-fading is used mostly for disco set-ups. It is the facility that enables an input such as a record-player turntable to be smoothly faded out and another faded in using the same control. It works something like a balance control on a stereo amplifier. At one extreme the one input is fully up and the other fully down. Turning it to the centre brings both in at equal levels, then at the other extreme the first input is fully down and the second fully up.

Priority enables one channel to reduce levels or silence all others automatically when it is opened, then restore them afterward. It is often used in factories where music may be carried by one channel, and announcements made on another. It could also be used for alarm signals that would have priority over both entertainment and announcement channels.

Monitoring

Many mixers have a small built-in amplifier with its own volume control to supply a pair of headphones. This is invaluable for checking quality, and determining whether any fault lies in the mixer or elsewhere.

It also permits another facility known as *pre-fade monitoring* to be used. This allows the monitoring of gram or tape inputs before they are applied to their respective faders so that they can be heard in the headphones with the faders turned down. The object of this is to enable the cueing of required sections of tape or disc before they are faded up through the system.

Visual monitoring of the mixer output is provided on most of the better mixers, though not on the more modest models. For serious public address it is essential. Output meters are used in most cases, though some have a barograph which is a row of light-emitting diode indicators that light up progressively as the signal increases. Calibration is in decibels with 0 dB representing the maximum undistorted output of the mixer. The range usually extends from −24 dB to +3 dB (Figure 40).

The output meter, often called a volume unit (VU) meter, is invaluable for keeping the peaks below the maximum level where overloading and distortion starts. Yet it also ensures that the level is high enough to provide sufficient drive to the power amplifiers and maintain a good signal-to-noise ratio. It further shows up level differences between channels and enables them to be balanced so that there are no distracting volume differences. Any fault affecting the signal level can also be immediately observed.

Output impedance

The output voltage and impedance differ considerably between various models. Voltages from 220 mV up to 2.5 V and impedance from 200 Ω up to 47 kΩ

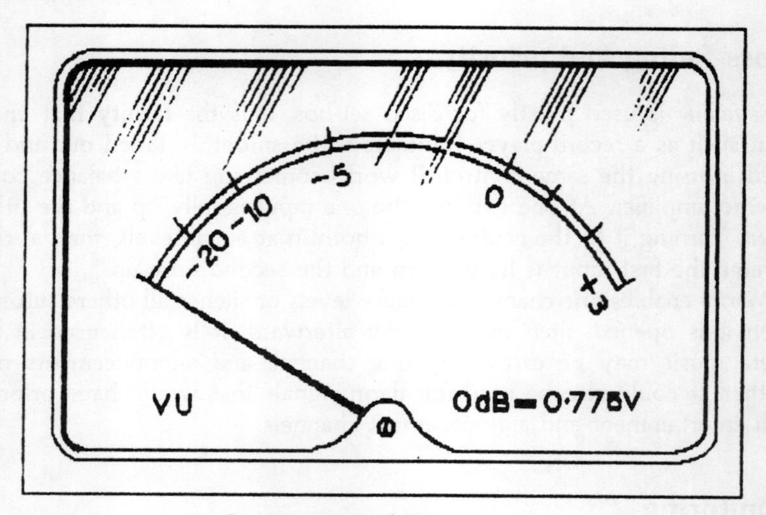

Figure 40 A typical VU meter.

are typical. The choice depends on the following power amplifiers and the cable length between mixer and amplifiers.

Impedance of the load should always be several times that of the source if it is not to reduce its output (see Chapter 3). Where a number of amplifiers are to be used, their input impedances will be in parallel across the mixer output and could present quite a low-impedance load.

The output voltage is dependent on the impedance, so if a high voltage is quoted, the impedance will probably also be high. Capacitance of the cable forms a shunt impedance at high frequencies. This is negligible for short lengths, but could be significant if the mixer is situated some way from the amplifiers. In such a case, a low mixer output impedance is essential to swamp the capacitance reactance, otherwise there will be a loss of high frequencies. To achieve this impedance an emitter-follower output stage is usually used.

Most professional mixers deliver 1 milliwatt output at an impedance to 600 Ω. From $\sqrt{(WZ)}$ where W is the output in watts and Z is the impedance, this gives an output voltage of 0.775 V. These are the standard output parameters, though as indicated earlier there are many variations from them. For general use with different temporary installations, a 600 Ω output mixer should be employed, as this will serve in most situations.

Master control

The master control enables the faders to be operated at a practical setting. If, for example, the gain is such that feedback starts at a low fader setting, all control of level has to be squeezed into the space between that setting and zero.

This makes the operation very critical and very easy to go over the top and start feedback.

By reducing the master control, feedback comes at a higher fader setting, so giving more space to operate in. If the master is turned too low, the faders never achieve the maximum level possible before feedback, so there is an optimum setting for it. Usually, a position that gives feedback with the faders about three quarters up is the best.

Power supplies

Most mixers are mains operated, but in view of the high gain available and the low signal levels of the microphone inputs special precautions must be taken to avoid hum.

The first of these is to minimize the possibility of magnetic pickup from the mains circuits, particularly the mains transformer. In all cases, a toroidal transformer should be used. These have a core shaped like a doughnut, and the circular configuration confines nearly all the flux to the core. With little stray field, the induction of hum into sensitive circuits is greatly reduced. Additionally, all circuits with low signal levels should be physically separated as far as possible from the transformer.

Wiring of all earth or negative leads should be to a single point to avoid common-impedance coupling, and in particular the reservoir capacitor should be earthed directly to the chassis and not via any wiring or point that is common to any other circuit. The negative connection of this component carries a high ripple current which will surely inject hum into any circuit it shares a common conductor with. Long signal-wiring, such as that to the faders, should be screened.

A suitable voltage regulator should be used in the d.c. supply line. This is not to regulate the d.c. voltage, which is not particularly critical, but the supply-borne ripple. A regulator will suppress ripple to a far greater extent than a large smoothing capacitor.

Volume compression

Volume compression can be built-in to a mixer or used as a separate unit. Few mixers offer compression as a standard option, but with integrated circuits that have most of the necessary circuitry, a compressor can be built with very few peripheral components, and fitted inside any mixer that has a small amount of internal space to spare.

The object is to automatically reduce loud volume peaks, to even up variations in level when an animated speaker ignores the microphone, and minimize the effect of explosive consonants and microphone thumping.

One integration circuit (I.C.) is the NE 570, and a circuit using it is shown in Figure 41. This is a dual channel unit, so two independent compressors could be built with it, although there is little point in this for public address. The

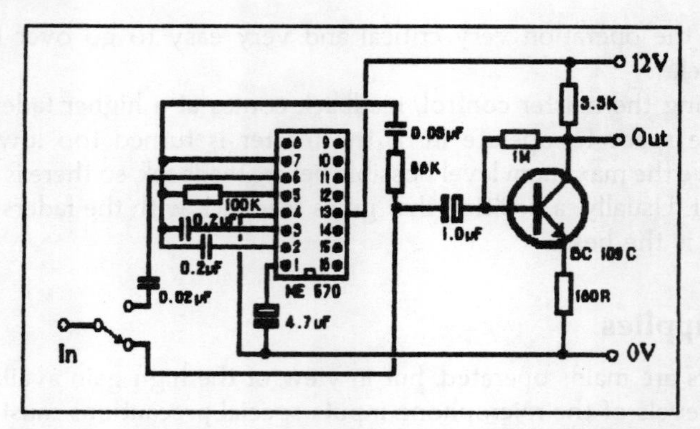

Figure 41 Compressor circuit. Input can be switched from compress to bypass. Compression starts above 50 mV.

device consists of a full-wave rectifier to detect the average signal level, a linearized temperature-compensated variable-gain cell which is current controlled, and an op amp.

The output of the rectifier is used to control the cell which is connected in the op-amp feedback circuit. As the output increases, so does the current through the cell, hence the amount of negative feedback. A 6 dB increase in the input produces only a 3 dB increase in output. Compression is non-linear, which means that there is virtually no compression of low-level signals, but compression increases with level, which is a desirable characteristic.

For 50 mV into input 1, the output is 34 mV, for 100 mV it is 52 mV. It thus needs to be connected just prior to the mixer output stage to operate over this non-linear area. If the signal input is much below 25 mV there is no compression. Input 2 is an uncompressed input with zero gain that can be switched in when the compressor is switched out.

Automatic channel switching

Automatic channel switching activated by sound can be useful where a number of inputs are used in random sequence, such as in a debating chamber or conference. If all channels are kept open much extraneous noise may be picked up and there is certainly greater probability of feedback. Manually fading up each channel as required is the usual alternative, but there is a high risk of delay in responding, or of fading in the wrong channel. Either way, part of the contribution is lost.

In one type of automatic control, the circuit checks the signal levels of all inputs and opens the channel having the highest level. This becomes the reference level and others with similar levels are opened as soon as they become active. If the signal drops well below the reference in any input it is switched off. Random noise, being much lower than the reference, does not actuate the circuit.

Individual automatic switching units for one channel have been available which switch on whenever sound over a certain level is received. These have a socket to receive the microphone plug, and a lead to plug into the mixer input socket. The principal disadvantage is that a unit is required for each channel which makes the set-up rather bulky and clumsy.

A circuit devised by the author is shown in Figure 42, and can be built on a $1\frac{1}{4} \times 2\frac{1}{2}$ in (31×64 mm) piece of matrix board. Room for several, one for each channel, can be found in many mixers.

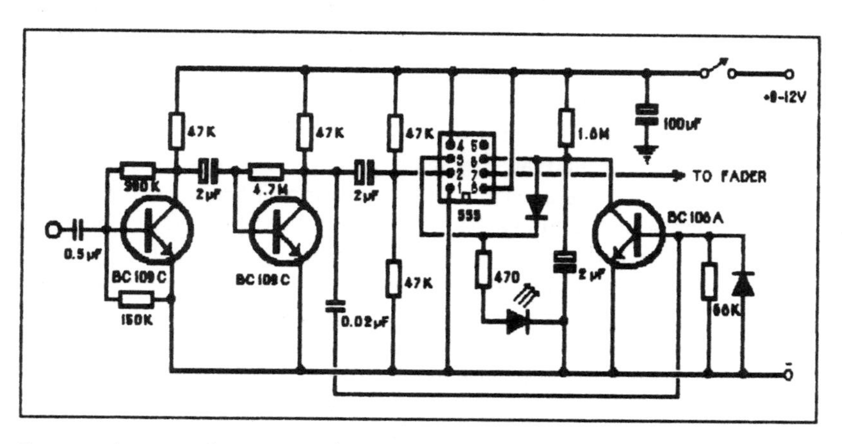

Figure 42 Auto-control circuit. Sound from the microphone opens the channel and illuminates the l.e.d. The channel mutes 5 seconds after sound ceases.

The circuit is designed around the 555 timer integrated circuit, which switches from one state to another when triggered by a square-wave pulse, then reverts to its original condition when a timing period has elapsed. This period is determined by an external resistor and capacitor.

The integrated circuit timing-capacitor discharge section is used to provide a short-circuit to earth, which is connected to the top of the appropriate channel fader, thereby muting it. Timing-capacitor discharge is effected by the timer output circuit via a diode which prevents it recharging when the output goes positive. A light-emitting diode is also connected to the output to give visual indication of when the channel is open.

The first few cycles of signal are amplified to the required level by the two-stage *R/C* coupled amplifier. High h_{fe} C-type transistors are used to get sufficient gain as the amplifier must be capable of triggering from a comparatively low input. The second stage is biased so that it overloads and so produces the square wave necessary for triggering the timer. The short is thereby lifted and the channel is opened.

Timing period is set for 5 seconds so that the channel mutes after 5 seconds of silence, so keeping it open during short pauses in speech. To prevent it from constantly switching on and off every 5 seconds, the capacitor is continually discharged by TR3 connected across it. This is kept conducting by a positive potential developed across a diode from the output of the amplifier. The capacitor thus remains discharged as long as signals are present, but starts charging when they cease.

TR3 must be of the low h_{fe} A-type in order to prevent it from conducting under the influence of low-level background noise. Also the 0.02 μF coupling capacitor must be of a low-leakage type to prevent the base from acquiring a permanent positive charge.

Sensitivity of the circuit can be controlled by varying the supply voltage with a series variable resistor. This has the advantage that the sensitivities of all the switches can be controlled at the same time by a single control. Optimum sensitivity depends on the acoustics of the location: if too sensitive, all channels switch on from sound coming back from the loudspeakers; if not sensitive enough, the first word or so is missed, especially with quiet users. As all microphones normally work in the same environment, optimum switching sensitivity is usually the same for all and a single control is convenient. It should be decoupled with a high-value capacitor at the circuit side.

Another refinement is the use of the CMOS 555. A slight click is sometimes heard when the circuit switches off, but the lower current taken by CMOS timer reduces switching noise to an almost imperceptible level.

Although automatic switches can be very useful in the right conditions they are not without disadvantages. As they are triggered by the first few cycles of signal these are lost before the device switches on. Although barely noticeable, clipped starts can be detected. With quiet users, the switch-on may be delayed further and the first syllable be missed. To minimize this effect, sensitivity should be as high as possible without ambient sound starting to switch the other channels. Sometimes, though, a hand microphone may be put down and left in a vulnerable position, such as facing the auditorium, and so the sensitivity of the whole switching system may have to be reduced to prevent this one from switching on.

One way that clipped starts could be avoided would be to use a bucket-brigade delay line to delay the input signal to the fader, so that the channel would be opened before it arrived there. The amount of delay would have to be limited if it were not to be noticeable to the audience, and this would reduce the amount of clipping that could be avoided. A delay line would be needed for each input although they could all be controlled be a single clock generator.

Really, the success of automatic switching depends on the level of ambient sound reaching the microphones: when it is low, high switching sensitivity can be employed and the system works well, but when ambience is high it may even have to be switched off and control revert to manual.

These are the principal features to be found on programme mixers. As stated before there is a wide variety from the simple to the highly complex. The illustrations in Figure 43 are just some examples of the range that is available.

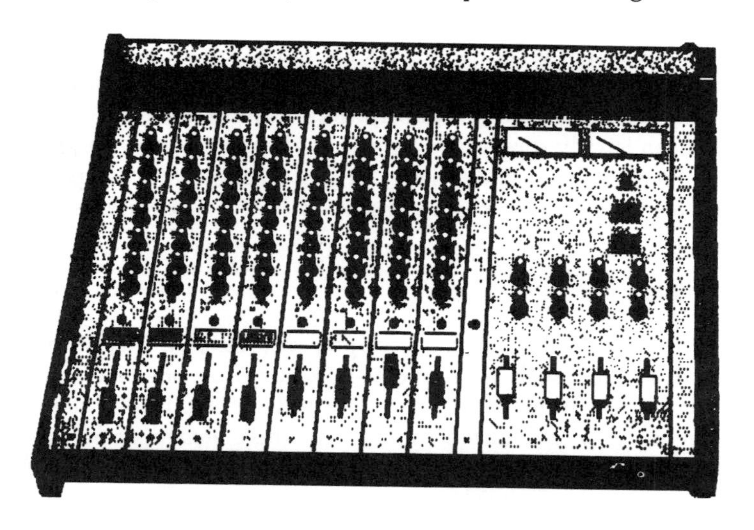

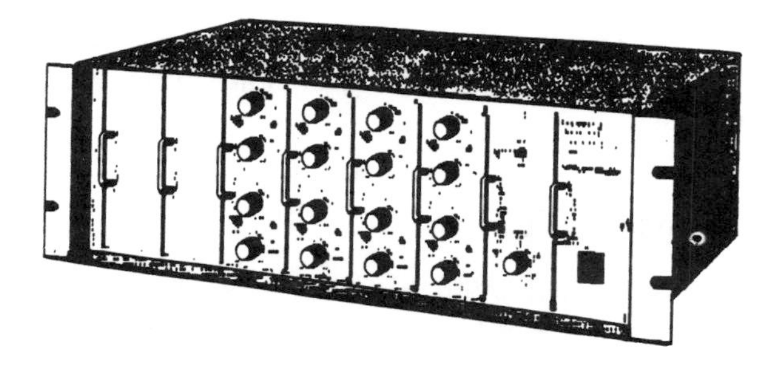

Figure 43 Various types of mixer.

7 Public address amplifiers

The building blocks of all modern amplifiers are transistors, whether discrete or as part of a complex integrated circuit. Since their first appearance many different basic types have been produced. Originally they were made from germanium, but these have been largely superseded by silicon. The semiconducting materials used are either P-type, having a deficiency of free electrons, or N-type, having a surplus. The main transistor types are discussed in the following sections.

Planar transistor

A slice of pure monocrystalline silicon is doped with either a P-type impurity, such as boron, or N-type, such as arsenic. It is oxidized on one side and a hole is etched in the oxide. Through this the opposite type impurity is diffused to form a well. The wafer is re-oxidized and a smaller hole etched. Impurity of the original type is diffused through this to form a smaller well within the previous one. Thus a PNP or NPN sandwich is formed. The centre 'filling' can be made very thin by careful control of diffusion, so large gains are obtainable.

A large number of units can be made simultaneously on the same slice, so manufacturing costs are considerably reduced. Furthermore, generated heat causes both wells to diffuse in the same direction, so reducing the possibility of short-circuiting.

Epitaxial planar transistor

This is similar to the planar type but a P-type layer is grown on to the surface of the N-type wafer before the oxide layer is deposited. This serves as the base, so only a single diffusing process is required to form the emitter. Most small transistors are of this type because they are easier to make, have higher breakdown voltage rating, higher gain, and slightly lower noise. They are made in both PNP and NPN versions.

Field-effect transistor (FET)

Various constructions are used. With one, P-type beads are either alloyed or diffused on either side of a slice of N-type material similar to a bipolar transistor. But instead of current passing from one to the other across the slice, it passes down the slice between the beads which control it by an electrostatic field. The material is termed the *channel*, and can be either N or P-type. The emitting electrode is the *source*; the control, the *gate*; and the final electrode, the *drain*. The device is thus voltage controlled and has a high input impedance.

MOSFET (metal-oxide semiconductor FET) As there is no need of direct contact between the gate and the channel, they are insulated in the MOSFET by a layer of silicon oxide. It thus has an even higher input impedance than the ordinary FET.
VMOSFET (V metal-oxide semiconductor FET) Power is limited when current passes along a narrow channel, so here it is made to pass *through* the slice like a bipolar transistor, thereby achieving a greater conduction area. The gate is formed into a V cut into one surface which produces a surrounding field in the material.

Output stages are increasingly using MOSFETs. Their negative temperature coefficient virtually eliminates the risk of thermal runaway. The absence of minority carriers in the control region gives a fast response, hence a good slew rate, and the high input impedance simplifies the driver stage.

A static voltage applied to the gate can puncture the oxide layer and so destroy a MOSFET device. Conductive packing materials are used to avoid this. The hand should be discharged to earth before handling, and soldering irons earthed when installing.

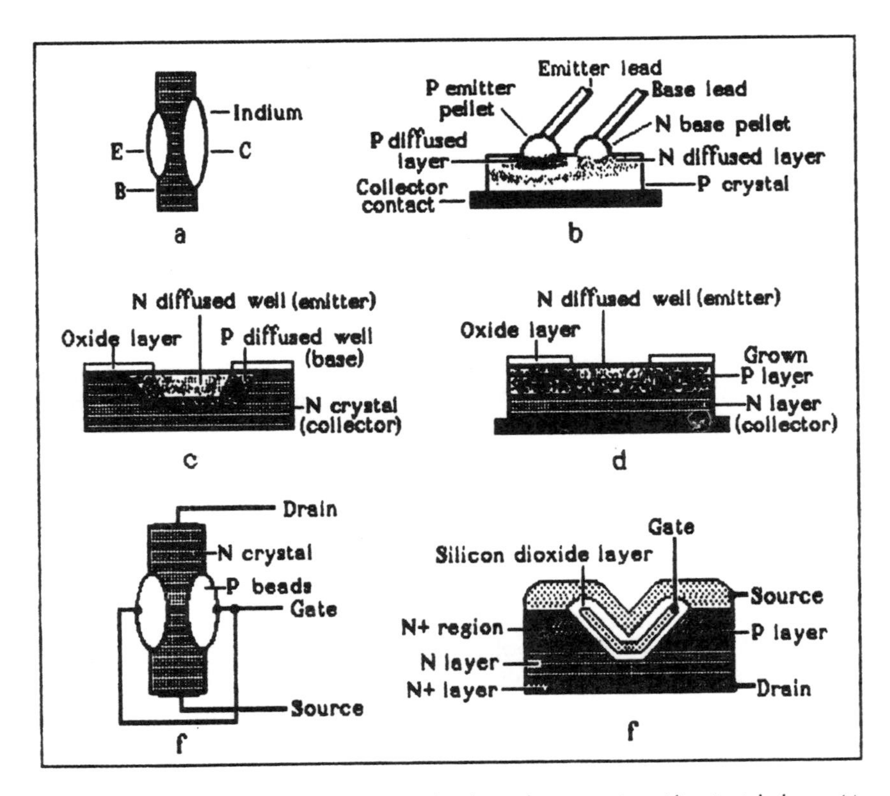

Figure 44 (a) Junction transistor; (b) alloy diffused; (c) planar transistor (d) epitaxial planar; (e) N channel FET; (f) VMOSFET.

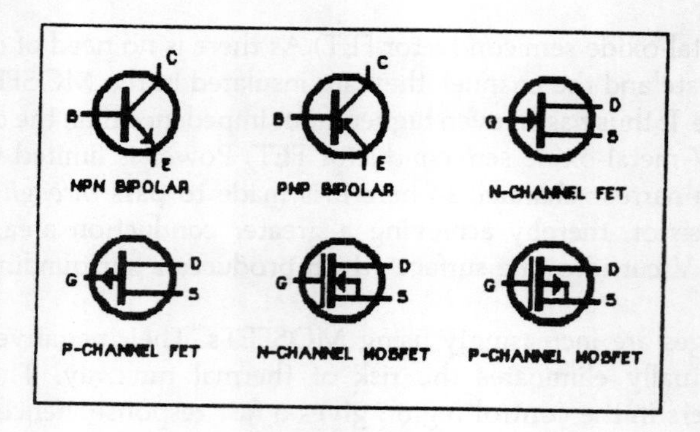

Figure 45 Transistor types and symbols.

The construction of these and their circuit symbols are shown in Figures 44, and 45.

Admittance Symbol Y. The small signal common-source transfer admittance for a FET, symbol Y_{fs}, denotes the drain current/gate volts. The unit is the *Siemens* which is equivalent to the mho. It is usually specified in μS or mS.

Alpha Symbol α. The current gain of a transistor in the common base mode, now rarely used.

Beta Symbol β. The current gain of a transistor in the common emitter mode, now superseded by h_{fe} for small signal, and h_{FE} for d.c. gain. The h_{FE} is obtained by dividing the collector current by the base current.

Bias. A bipolar transistor must be forward biased in order to conduct, that is a positive current must be injected into the base for a NPN, or a negative current for a PNP device. With FETs of the enhancement type, forward gate voltage bias is required (positive with N-type and negative with P-type channels). With depletion type, reverse gate voltage bias is necessary. Most output devices are of the enhancement type and so require bias of the same polarity as the supply, thus simplifying the biasing arrangements.

Heat dissipation

A major factor, especially with bipolar transistors. A temperature rise produced by the power expended in the device, reduces the resistance so increasing the current and power, thereby raising the temperature further. Current and temperature continues to rise in what is called *thermal runaway,* until the device short-circuits and is destroyed.

Temperature must therefore be controlled by adequate heat removal. There are three barriers it must cross: (1) the thermal resistance between the collector junction and the casing of the device; (2) the thermal resistance between the device and its mounting; and (3) the resistance between the mounting which is usually a heat sink, and free air. If the device is free standing, (2) and (3) are combined.

Thermal resistance is defined as the temperature difference that exists across the barrier for each watt of power produced, and is specified in °C/W. Maximum ambient temperature, heat-sink resistance or power can be determined given the others, and the device junction maximum temperature. The relationship is.

$$T_a \leq T_j - P(\Phi_j + \Phi_c + \Phi_H); \quad \text{or} \quad \Phi_H \leq \frac{T_j - T_a}{P} - (\Phi_j + \Phi_c);$$

$$\text{or} \quad P \leq \frac{T_j - T_a}{\Phi_j + \Phi_c + \Phi_H}$$

where T_a is the ambient temperature; T_j is the junction temperature; P is the power in watts; and Φ_j, Φ_c, Φ_H, are the thermal resistances of the junction-to-casing, casing-to-mounting, and heat-sink respectively.

Maximum junction temperatures are between 150 and 200 °C for silicon and 100°C for germanium transistors. The following are approximate values of Φ_j for various encapsulations, but exact value depends on type:

TO3	0.5–2.5°C/W
TOP3	1.5°C/W
TO5	15–35°C/W
TO39	15–36°C/W
TO66	4.5°C/W
TO220	3–4°C/W

Mica washer resistance between TO3 devices and heat-sink is 0.5°C/W, while direct contact between device and mounting is about 0.2°C/W.

H_{FE}/H_{fe}

The d.c. current gain, and small signal current gain. Formerly called beta. Large spreads are common up to 700%, so some types are divided into three groups having a suffix A, B, or C, with C having the highest gain. Sometimes these are colour coded with a red (A), green (B), or blue (C) spot. The H_{fe} is usually specified at a single operating frequency which for audio is 1 KHz; it is much less at frequencies near its upper limit (100 KHz for audio). Gain also varies with collector current, being lower at the higher currents and often also at very low currents.

Limiting parameters

Data sheets specify maximum voltages and currents for the various electrodes. These are usually for 25 °C (77 °F), and must be derated for higher ambient temperatures or smaller heat-sinks. Uprating is also possible if larger heat sinks are used which can be calculated from the above formula. For long life and reliability a device should be operated well within its maximum ratings.

Basic circuit configurations

Transistors can be operated in a number of circuit configurations to give a variety of characteristics. Some of these, which are illustrated in Figure 46, are as follows:

Common Base. The signal is applied across the emitter–base junction, the base being grounded and common to input and output circuits. Output is taken from the collector side of the load. Has voltage gain but current gain is less than unity. Lowest input impedance of about 50 Ω, highest output impedance of around 1 MΩ. O° phase shift.
Common Emitter. The signal is applied across the base–emitter junction, the emitter being grounded and common to input and output circuits, output is from the collector side of the load. Has voltage and current gain. Input impedance 1–2 kΩ, output impedance about 30 kΩ 180° phase shift. This is the most common configuration
Common Collector. Also called *emitter follower.* Signal is applied across base to ground. Output taken from the emitter side of the emitter load. The collector is grounded as respects the signal (but not necessarily the supply), and so is common to input and output. Has current gain but voltage gain is less than unity, being

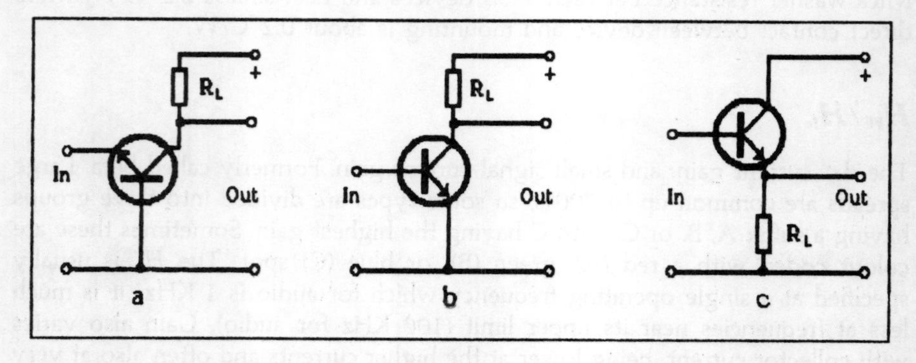

Figure 46 (a) Common base; (b) common emitter; (c) common collector.

$$V = \frac{Z_e}{Z_e + \left(\dfrac{1}{g_m} + \dfrac{Z_s}{H_{fe}}\right)}$$

In which Z_e is emitter impedance, Z_s is source impedance; g_m is the mutual conductance, which depends on internal resistances and emitter current, typically 50 at 1 mA for small transistors, very low for output transistors.

High input impedance is a feature, being approximately equal to the H_{FE} multiplied by the emitter load, shunted by the base bias resistor if used. Output impedance can be very low. There is 0° phase shift.

Darlington pair/super-alpha pair. Two transistors have their collectors connected, and the emitter of the first feeds into the base of the second. The input is taken to the base of the first, and the output from the collector or emitter of the second. The first unit is an emitter follower and a very high imput impedance in excess of 1 MΩ is achieved. The gain is equivalent to the individual H_{fe}s multiplied. As power output transistors have low gain, a Darlington output pair in a single encapsulation is often used to give gain with power, (Figure 47(a)).

Differential pair/long-tailed pair. Two transistors are connected with a common emitter resistor which stabilises their d.c. operating conditions. Input is connected across the two bases and the output across the two collectors. If one side of the input circuit is earthed, one of the bases can be a.c. coupled to earth. Signals of opposite phase are produced at the collectors so when these are combined, harmonic distortion generated by transfer curvature is cancelled. Supply ripple or noise affect both transistors and so is cancelled. This is termed *common-mode rejection.* Differential pairs are often used for amplifier input circuits. (Figure 47(b)).

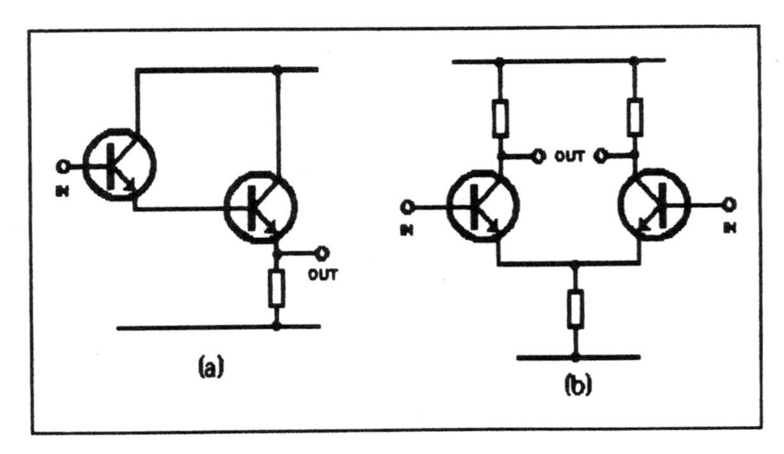

Figure 47 (a) Darlington pair; (b) differential pair.

Base bias

In order that both half-cycles of the applied signal can be amplified equally and without distortion, the transistor must be biased to operate at the mid-point the straight portion of the transfer characteristic. Excursions of the collector current either side of this point thereby remain in the linear portion. This is termed *Class A* operation. If the bias point is not at the centre, one half-cycle encounters a curve earlier than the other, so the signal-handling capacity is reduced to that of the shorter section.

For single, common emitter transistors, the device is mid-point biased when the collector voltage is roughly half that of the supply, or the emitter voltage for the common collector mode. This is achieved for the common emitter stage when the value of a base bias resistor, which is returned to the collector, is equal to the collector load resistor multiplied by the H_{FE}.

Bias can be derived from a tap on a potential divider across the supply, or from a single resistor to the supply rail or collector. The latter gives negative feedback over the stage which reduces gain, noise and distortion. With directly coupled circuits, the bias is obtained via the previous transistor.

A resistor is usually included in the emitter circuit to provide thermal stability and also to even out the effects of H_{FE} spreads. A device with a higher H_{FE} than designed for has a higher collector current for the given bias. The result is a larger voltage drop over the emitter resistor which reduces the base–emitter voltage and with it the collector current. The resistor is bypassed with a high-value capacitor to avoid negative feedback of the signal, unless it is required.

Interstage coupling

Coupling from one stage to the next can be by means of a capacitor between the collector and base of the following stage. Each stage can thus make full use of the available supply voltage, the d.c. voltage of one does not upset that of the next should a fault develop, and it is easy to diagnose faults by meter readings.

Alternatively, the stages can be directly coupled by using both PNP and NPN transistors. Phase shift over the stages is reduced because of eliminating the reactive components, important when heavy negative feedback is applied as it is with hi-fi amplifiers, but less so with public-address types. However, the supply voltage must be shared, a fault in an early stage can cause catastrophic failure in later ones, it is impossible to diagnose such faults with a meter, but it is cheaper to manufacture. So the direct-coupled circuit is most commonly used.

Output circuits

For output circuits, class A is only about 20% efficient, which means that for large output powers, considerable heat must be dissipated. The *push–pull* circuit using two transistors is universal for all but very small output stages. In this, the current increases through one transistor as it decreases in the other, both being biased to their mid-points. Cancellation of second harmonic distortion is thereby achieved and about $2\frac{1}{2}$ times the power of a single transistor is obtained.

Because of its poor efficiency and high heat generation, class A is rarely used. An alternative, possible with the push–pull arrangement, is class B in which the transistors are zero biased, and only a small standing current flows through them. Each unit then handles only one half-cycle. However, there is a displacement where the two halves join that generates *cross-over distortion,* which is mostly third harmonic. As the discontinuity is proportionally larger for small signal amplitudes than large ones, the audible effect is worse at low volumes.

A related effect is *switching distortion.* As each transistor switches off it stores a charge and the corners of the wave are rounded off at the switching point, so that the two halves do not connect.

To reduce these distortions a certain amount of bias is applied so that the operation is part way between class A and class B. Negative feedback helps reduce them further. Heat is produced by the standing current which must be dissipated by heat sinks of suitable proportions.

With conventional push–pull circuits the phase of the signal applied to the base of one transistor must be opposite to that of the other so that the current through one increases as that through the other diminishes. A phase-splitter must thus be incorporated into the preceding driver stage.

The principle is illustrated in Figure 48. The transformer T1 has a split secondary winding which applies signals of opposite phase to the bases of the output pair.

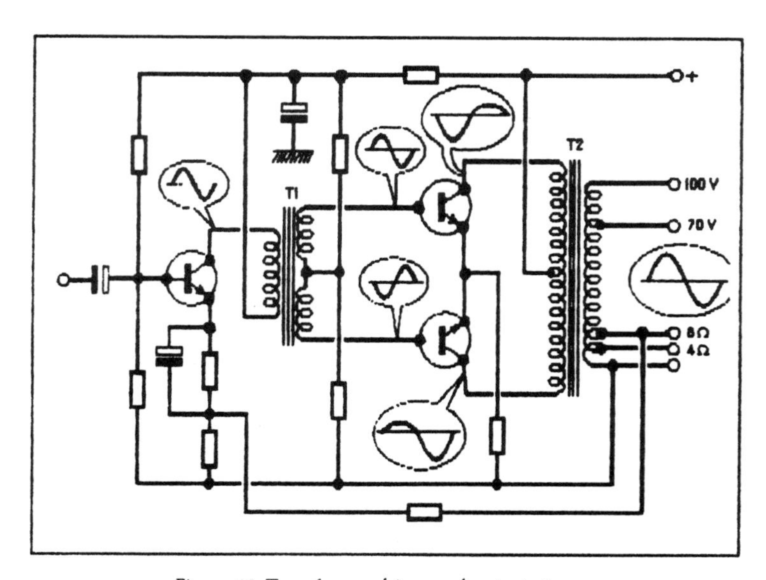

Figure 48 Transformer driver and output stages.

Signals of opposite polarity appear at their collectors, but the positive half of the wave in each case is partly suppressed by the class B biasing. These are re-combined in the split-primary output transformer so that an in-phase output appears at the secondary.

Transformers are bulky, weighty, cause phase shifts and are expensive, so a transformerless circuit is now almost always used. One version of this is the series *complementary* output stage, used for low or medium powers, employing a PNP and NPN transistor as the output pair. The first advantage of this circuit is that a negative-going signal turns on a PNP device whereas it turns off one of the NPN type. So, there is no need for phase-splitting and the two bases can be driven from the same source. The second is that the output signal appears at the junction of the two output emitters so there is no need for a split-primary transformer; a low-impedance loudspeaker can be driven directly from this point via a coupling capacitor.

In the case of a public-address amplifier a transformer is needed to provide the high impedance 100 V output which we will discuss in a later chapter. The primary of this is connected in place of the loudspeaker. However, this does not nullify the advantage of the series pair. The transformer does not require a split primary, and in particular the primary does not carry the output transistor standing current. It thus needs smaller gauge primary windings, and a smaller core than would be required for the transformer of Figure 48.

The operating principle of the complementary output stage is shown in Figure 49. At (a), a positive-going signal at the driver base produces a negative one at the bases of the output pair. This turns TR1 nearly off, but turns TR2 on. Current thus passes through it to charge the capacitor through the loudspeaker which makes a positive excursion.

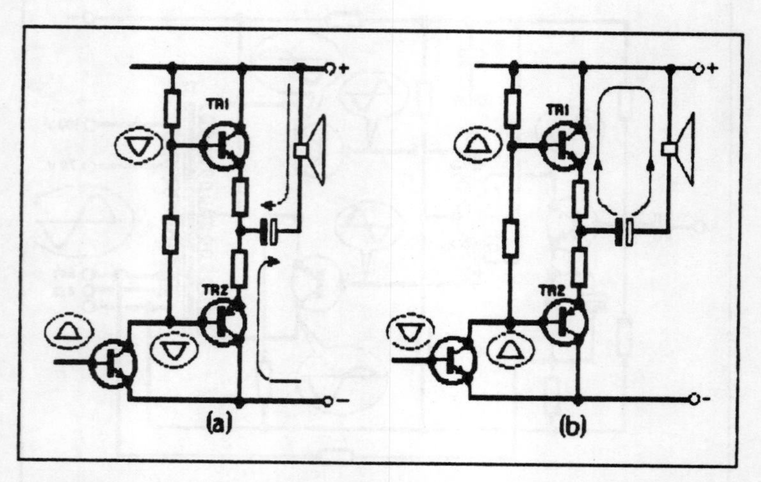

Figure 49 (a) Negative half-cycle turns on TR_2 and C charges through the speaker. (b) Positive half-cycles turns on TR_1 so discharging C through the speaker.

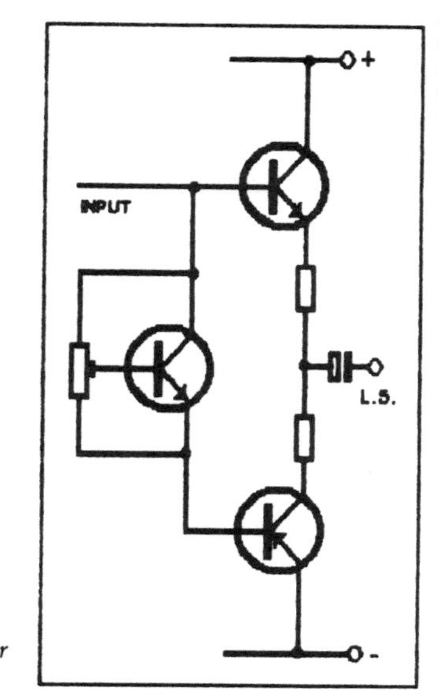

Figure 50 Temperature compensation with bias regulating transistor mounted on the heat sink.

Next, at (b), a negative-going signal at the driver produces a positive one at the bases which turns TR2 off and TR1 on. The capacitor now discharges through TR1 and the loudspeaker, which makes a negative excursion. The function of the capacitor is thus not just to isolate the loudspeaker from the d.c. potential, but it performs an essential part of the operation.

The resistor included between the two bases of the output transistors has an important function. It is part of a potential divider from the positive supply to the collector of the driver transistor, and so establishes a voltage difference between the bases. As the emitters are virtually at the same mid-point potential, the voltage biases and bases either side of this and so controls the current through the transistors.

It can take the form of a preset variable resistor which is adjusted to set the output current at the required level as indicated by a meter inserted in series with the output pair. The higher the resistance, the greater the current.

To reduce the possibility of thermal runaway in the output transistors, a transistor is often used in place of the resistor, with its current controlled by a preset resistor as shown in Figure 50. This transistor is mounted on the output heat sink, so that when the temperature rises, its resistance decreases, thereby reducing the bias on the output pair. The preset controls the output current as before. Alternatively a thermistor may be used for temperature compensation.

Matching

The output pair should be matched in H_{fe} as close as possible. Mismatched transistors produce unequal half-cycles and so generate second harmonic distortion. The amount of distortion depends on the ratio between the two H_{fe}s. Table 6 gives the values.

Table 6

Ratio H_{fe1}/H_{fe2}	1.1	1.2	1.3	1.4	1.5	1.6	1.7	1.8	1.9	2.0
Distortion %	2.4	4.6	6.5	8.3	9.8	11.5	13.0	14.5	15.5	16.5

These figures are not allowing for negative feedback which reduces the distortion according to the amount of feedback used.

When replacing output transistors it is not always easy to get a matched pair, but some suppliers will match them Alternatively it is a good practice to obtain four or more and match a pair with a transistor H_{fe} meter. The others can then be retained for spares.

The driver

The drive current required by an output stage is the output current divided by the H_{FE} of the output transistors. As this is usually well below 100, the drive current is quite large so the drive transistors must be power types with heat-sink cooling. The output impedance should be high to obtain a high ratio to the input resistance of the following stage. Then variations of its input resistance will be swamped and have little effect on the driver and the applied signal.

The driver output is directly coupled to the bases of the output transistors. It may consist of a single transistor, but more usually a Darlington coupling with each output transistor to provide sufficient voltage swing, because of the low H_{FE} of high-current output devices.

High-current PNP output transistors are not entirely satisfactory, so for high outputs it is the drivers that are usually configured into a complementary pair, thus feeding opposite phase signals to a series output pair that are both NPN (Figure 51).

MOSFET output stages

Increasing use is being made of the MOSFET device for output stages in place of the conventional bipolar transistor. Its principal advantage is that it has a negative temperature coefficient, that is the current decreases as the temperature rises, so it is not subject to thermal runaway. This confers another advantage in that no temperature compensation is required because it is inherent. So a simple resistor or preset can be used for bias between the bases as in Figure 52.

Another advantage is that MOSFETs are voltage controlled, rather than current controlled like bipolar transistors. The driver stage thus does not need to be a power transistor with a heat sink, and furthermore can usually provide sufficient

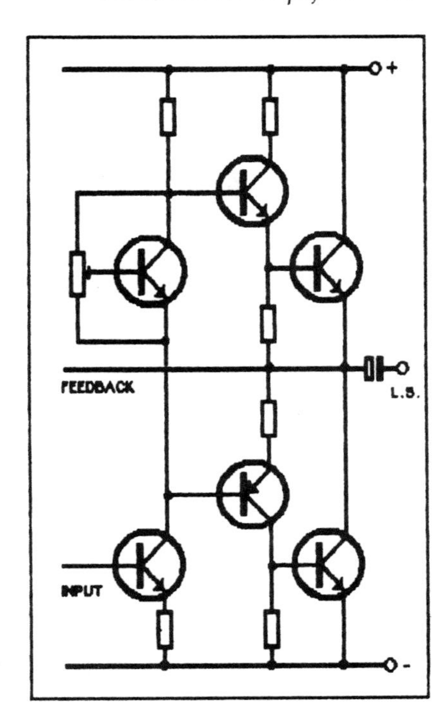

Figure 51 An NPN output pair driven by a complementary driver.

gain to drive the output without recourse to a Darlington pair. Both driver and output stage can thus be greatly simplified.

When high powers are required MOSFETs can be simply connected in parallel to give or even four or even more in the output stage. This cannot be done with bipolar transistors because it is virtually impossible to find a perfectly matched pair to run in parallel. If not matched, one draws a higher current than the other, gets hotter and so draws even more current. The result is that one ends up doing all the work while the other does very little.

With MOSFETs, when one draws more current than the other and gets hotter, its current is thereby reduced, while the cooler one takes more current. As this one gets hotter its current decreases. Both thus keep stable operating currents controlled by their own temperatures.

MOSFETs have a far higher switching speed of up to 100 times faster than bipolar transistors. This gives them excellent high-frequency response with very little phase shift, and so large amounts of negative feedback can be used while maintaining stability. This is useful for hi-fi amplifiers for which low distortion specifications are important, but less so for public address.

When first introduced MOSFETs were very expensive and so not often used, although the simplification of the circuit partly compensated. Nowadays the price differential is much less and they are frequently encountered in amplifiers.

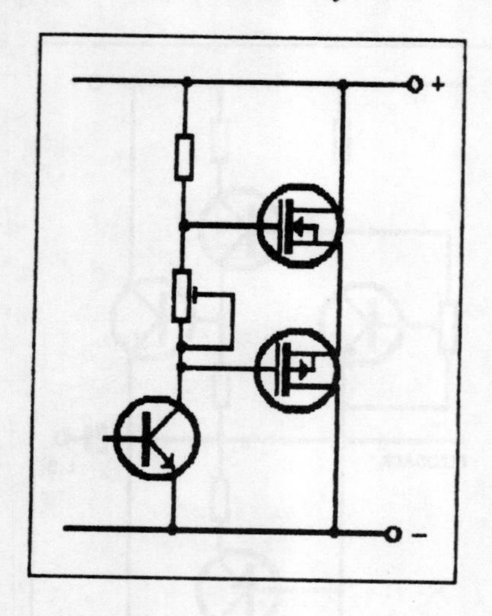

Figure 52 MOSFET output stage.

The one disadvantage is that they can be damaged by a high voltage spike to the gate. This could be produced by high transient noise inadvertently applied to the amplifier input.

Negative feedback

If a fraction of the output of one or more stages is fed back out of phase to the input a gain reduction occurs. Also harmonic or frequency distortion generated within the feedback loop is reduced. As the fed-back signal is re-amplified and fed back again, the final result is the sum of an infinite series, but this can be calculated from:

$$g = \frac{G \times n}{G + n} \quad \text{and} \quad d = \frac{D \times n}{G + n}$$

in which g is the final gain. G is the gain without feedback; n is the fraction of the output signal fed back; d is the final distortion and D is the distortion without feedback.

Large amounts of negative feedback are used in modern hi-fi amplifiers. Stages are designed with greatly increased gain to compensate for the loss due to feedback. Although it is rapid, feedback is not instantaneous, so if a sudden transient signal occurs, it can cause overloading in the brief interval before the feedback arrives to throttle back the gain of the controlled stage. The effect is termed *transient intermodulation distortion.*

The loudspeaker generates back e.m.f. which appears at the amplifier output and so is fed with negative feedback to an earlier stage. It is delayed due to cone inertia and contains distortions produced by the non-linear cone response as well as reactive elements in the cross-over network. As back e.m.f, is in opposite phase to the original signal, it becomes a positive feedback. Distortion is thus produced which is not evident when it is measured with a resistive load.

Some designs apply feedback in several loops over individual stages instead of a single loop over several stages. The trend now is to regard feedback as a necessary evil and to reduce it to the minimum.

With public address amplifiers the amount of feedback is modest and many of these problems are avoided. There is one situation though that could give trouble. Normally the load consisting of transformers and loudspeakers is inductive, but long loudspeaker feeders introduce capacitance. This could be sufficient to cause phase shift sufficient to turn the negative feedback positive at high frequencies. It may occur at frequencies higher than human audibility, but the resulting oscillation could overload the output stage, producing distortion and possible thermal runaway. High-frequency response is usually limited to avoid this.

With an installation that involves very long loudspeaker feeders, it is advisable to check with the suppliers of the amplifier whether it is stable with capacitive loads. If there is any doubt with a particular model it can be checked by connecting an oscilloscope across its output terminals, then capacitors shunted across the output starting at 0.01 μF, and progressing at 0.01 μF steps up to 0.1 μF. No spurious waveforms should appear on the scope.

Power supplies

The power supply circuit has many forms, but those used for amplifiers comprise just a few basic types consisting of three essential components, the mains transformer, the rectifier and the reservoir capacitor.

The transformer primary is designed for 240 V mains voltage, and the secondary for the required voltage, the turns ratio being proportional to the voltage ratio. For full-wave circuits and split polarity bridge circuits, the secondary is centre-tapped. A transformer is rated in VA, which is the product of the secondary volts and amperes, and has a regulation rating which specifies the percentage difference between the off-load voltage (V_{OL}) and the full-load voltage (V_{FL}). The formula is

$$\text{regulation} = \frac{V_{OL} - V_{FL}}{V_{OL}} \times 100\%$$

Toroidal transformers have very low flux leakage, hence a minimum hum field, and so are preferred for audio equipment.

Silicon rectifiers are now universally used because of their small size and high current ratings. Voltage drop at low current is 0.6 A. When non-conducting, the combined transformer and reservoir capacitor voltage are effectively in series, and produce a *peak inverse voltage* of up to three times the applied r.m.s. a.c. voltage. This PIV must be considered when replacing a rectifier.

The reservoir capacitor charges from the applied voltage peaks and discharges into the load when the voltage falls. If the capacitance is large enough the reduction in charge and capacitor voltage is small before the next peak arrives to top it up. Thus the d.c. level is maintained. The rise and fall in voltage produces a ripple on the supply line at the mains frequency for half-wave rectification, and twice the mains frequency for full-wave. This can generate hum, so the ripple must be kept as small as possible.

The amplitude of the ripple depends on the current as the discharge hence the slope, is greater with higher loads (Figure 53). Increasing the capacitance reduces the ripple, but it also increases the charging current through the rectifiers on peaks, as they top up the capacitor in a shorter time. The charging current on switch-on is also greater. Higher capacitance thus needs higher current-rated rectifiers and the provision of small-value series resistors as surge limiters.

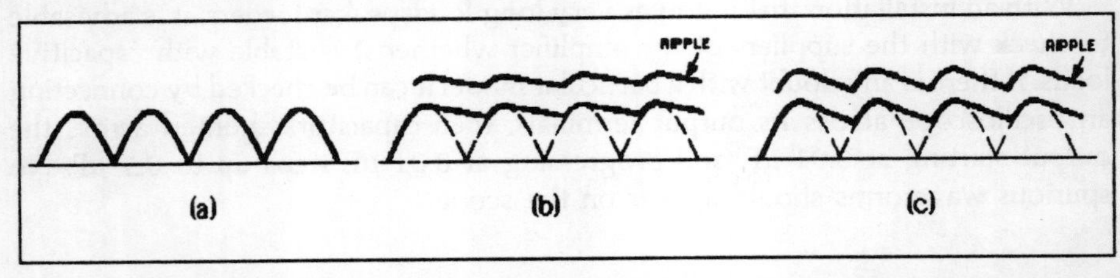

Figure 53 (a) Output of full-wave rectifier. (b) Reservoir capacitor charges during positive peaks and discharges during gaps so producing a ripple. (c) If capacitance is low or load current high, the discharge is steeper so producing a more pronounced ripple.

The ripple current, which is equal to the total load current, is following in and out of the capacitor continuously. It must therefore be designed to withstand this current, which is why reservoir capacitors are physically large. Any replacement must be chosen with a ripple current rating higher than the load current at full power.

Ripple reduction in power units supplying class A circuits can be achieved by a simple hum filter consisting of a series inductor or resistor and a further high-value capacitor termed the *smoothing capacitor*. Class B circuits produce large swings in supply current which give rise to corresponding voltage drops over any series resistance. The supply is thus modulated by the signal and severe distortion is generated in the output stage.

The *RC* hum filter thus cannot be used for a class B amplifier output stage, but can be employed for the supplies to earlier ones. This is fortunate, as it is the earlier stages that are most vulnerable to hum because of the large amount of following amplification.

Voltage regulators are often used to supply low signal-level input circuits as these give a d.c. output that is very stable and free from fluctuations such as hum ripple.

Half-wave rectifier

The simplest power circuit, it consists of a single rectifier in series with the transformer secondary and the load, with the reservoir across the load. D.c. output voltage is virtually that of the a.c. peak which charges the reservoir, so it is 1.4 times the r.m.s. rating of the secondary. The d.c. current is drawn during the whole of the suppressed negative half-cycle so a much larger current must be supplied by the transformer during topping up. Available d.c. is 0.28 times the a.c. current. The ripple is large, requiring a large reservoir, and more smoothing.

Full-wave rectifier

A centre-tap on the transformer secondary is usually connected to earth. Two rectifiers are wired to the ends of the transformer winding, and their cathodes (for a positive supply) are both connected to the supply line and the reservoir capacitor. Only half of the secondary is conducting at one time, so the d.c. voltage is half of that of the half-wave circuit, or 0.7 the total r.m.s a.c. voltage. D.c. current is equal to the a.c.

Bridge rectifier

Four rectifiers are used in a bridge-type circuit, with the transformer secondary connected to the anode/cathode junctions of both pairs, while the d.c. supply line negative goes to the two free anodes, and the positive to the free cathodes. It is easy to forget which way the diodes are connected in a bridge circuit; as a mnemonic remember the N in negative and anode. The two aNodes go to the Negative, therefore two cathodes must go to the positive; the junctions of each then go to each end of the transformer winding. An alternative is to consider the appearance of the diode, some have a pointed end, others have a line around them at one end. So remember: *Lines and points are positive.*

The reservoir capacitor is connected across the positive and negative of the supply. The whole transformer winding is used each half-cycle, so the d.c. voltage is 1.4 times the rms a.c. rating. Current is 0.6 of that taken from the transformer. So for the small cost of two extra diodes, the transformer winding does not need a centre-tap, and the number of secondary turns is halved, although the wire gauge must be thicker to supply the extra current. There is thus a net saving, and so the bridge has virtually taken the place of the full-wave circuit.

Split-polarity full wave

A separate positive and negative supply line with a common earth is frequently required for many op-amps and output stages. This is achieved with what looks like a bridge circuit but is actually a double full-wave circuit. Four diodes are connected in the same way as for a bridge across the transformer secondary, but the winding has a centre-tap which is earthed. The diodes are thus two pairs of full-wave rectifiers connected for opposite polarity. Voltage for each polarity circuit is the same as for a normal full-wave circuit, 0.7 of the a.c. supply. Current is theoretically half of the a.c. supply for each circuit, but if the load is mainly a class B output stage, one section will be supplying high current when the other is passing little. The current rating requirement can thus be reduced.

Output power

Amplifiers are classified mainly by their power output. These normally range from 20 to 250 W, but there are larger ones. For large installations, though, it is better to have several amplifiers of modest output power than a single high output one. This gives versatility: levels can be independently controlled, different programmes can be relayed to different areas, and if there is a breakdown, only part of the system is affected.

Power is sometimes confused with gain or amplification. If there is insufficient volume, it is assumed that a higher-powered amplifier is required. This is rarely if ever the case. A more likely reason is that a low-level microphone is being used with an input of insufficient sensitivity, or most likely, feedback is preventing the available gain being used.

Most public-address amplifiers except battery-operated ones range from 50 to 100 W, which is more than adequate for most systems other than large auditoriums and factories. The power rating is the amount of power the amplifier will deliver at the stated load without clipping and overloading. The main symptoms of an underpowered amplifier with a given load is distortion, although distortion can be caused by many other things. It is also likely to overheat the output transistors with possible thermal runaways unless they are MOSFETs.

The power available depends on the load. For low-impedance operation the higher the total load impedance, the smaller the power delivered. For example, one 40 W amplifier with a rated output impedance of 4 Ω produces 34 W at 6 Ω, 25 W at 8.0 Ω and 14 W at 16 Ω.

With 100 V line output (to be described in a later chapter), the output produced is the sum of all the transformer tappings connected to the line.

Phasing

Many public-address contractors have a number of amplifiers in their stock of equipment, these being needed for simultaneous jobs, standby operation and

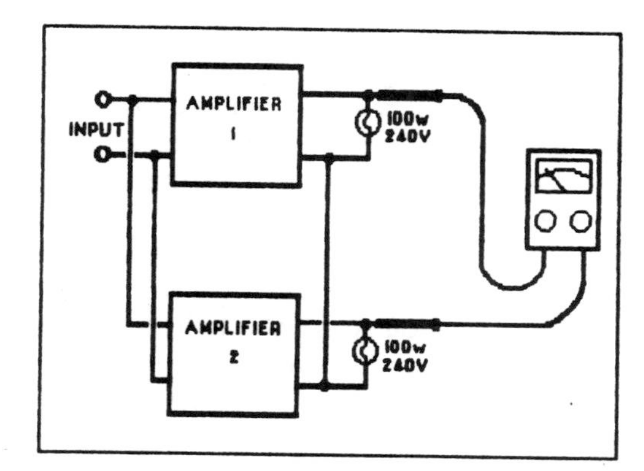

Figure 54 Phasing amplifiers. If a null cannot be obtained with both amplifiers operating they are out of phase.

large jobs needing more than one. It is an advantage to have all the same model as then fewer spares are required and familiarity with one model makes servicing easier.

However, it is likely that the amplifiers will have been acquired over a period from different sources and so constitute a rather diverse collection. In this case they should be phased to the same standard. That is the polarity of all the outputs should be the same for a given polarity input. If then two or more are used on the same job, the phasing of the loudspeaker system will be correct (more of this in a later chapter).

Phasing can be done quite easily by connecting an audio oscillator simultaneously to the inputs of two amplifiers, the earthy terminals being connected together, and connecting a dummy load of a 100 W 240 V lamp across each 100 V output. Link the common loudspeaker terminals of both amplifiers and feed in a signal so that the lamps are just alight.

Now connect an a.c. voltmeter across the two 100 V outputs and adjust the gain of one amplifier to get a zero reading. If such is obtained then the amplifiers are in phase. If it is impossible to get a zero without turning both amplifiers right down, they are out of phase (Figure 54).

A similar test can be made on the low-impedance output if it is taken directly from the output transistors and not via a tapping on the transformer.

If the 100 V outputs are not in phase, the primary windings of one transformer can be reversed. If the direct low-impedance output is out of phase little can be done other than marking the output terminals accordingly. Other amplifiers can then be tested and made to conform to the originals.

An alternative method of checking the phase without using instruments is to connect a couple of loudspeakers that are of the same phase, one to each amplifier, and feed in some music from a tape. Listen with the loudspeakers well apart, then stand them side by side. If there is no difference in the sound quality, the amplifiers are in phase; if the bass disappears, they are out of phase.

Portable amplifiers

Portable amplifiers are normally designed to run from a 12 V battery, whereas the d.c. supply of mains-driven equipment is usually 48–60 V. Power is thus restricted, and the maximum for this voltage is around 50 W. MOSFETs cannot be used because they need several volts bias between source and drain, which is doubled with a complementary pair. This must be deducted from the supply when calculating maximum signal voltage swing, which would significantly reduce a 12 V source. Bipolar transistors have a base-to-emitter voltage of 0.6 V, which means a reduction of the supply of only 1.2 V.

Battery drain is not great, as class B circuits take high current only at maximum signal levels. At lower levels the current is in proportion, and at pauses, it is quite low. A 50 W amplifier would take about 5 A at maximum signal peaks, but about 2 A average. A fully charged 60 ampere/hour battery should thus last for 25–30 hours. This gives some idea of what is needed when planning outdoor events, although the drain will depend a lot on the programme material. It is of course always prudent to have a fully charged standby battery.

Some small amplifiers are designed for either mains or battery operation, and these can be very versatile. In some country areas that are prone to power cuts in stormy weather these could well prove their worth.

Battery amplifiers usually have a limited input capability of only one or perhaps two microphones.

Slave/mixer-amplifiers

A slave amplifier is one that has no low-level inputs. it just has a single input socket to accept the standard 0.775 V. Output facilities and meters are the same as for other amplifiers. It is intended to work from a mixer and so has only a single volume control. It may have tone controls.

Slave amplifiers sometimes have in–out sockets; these are connected in parallel, the 'in' receiving the lead from the mixer and the 'out' enabling another slave amplifier to be plugged to it. A number of amplifiers can thus be connected in a chain, limited only by their combined input impedances not falling below the mixer output impedance.

Mixer amplifiers have built-in mixers normally of two or three microphone inputs and a line input. This often restricts their use, and a mixer is frequently added later. This is wasteful, as the mixer section of the amplifier is then no longer used. It is better to plan for possible future requirements in advance and install a slave amplifier with mixer from the start, unless it is certain that the limited facilities of a mixer amplifier will be adequate.

8 Loudspeakers for public address

The loudspeaker drive unit is a familiar enough object, but the details of its construction and behaviour may be less well known. It is necessary that its principles should be well understood so that effective use can be made of it in public address installations. A cross-section of a typical unit is shown in Figure 55.

Cone surround

The cone is usually made of paper and is fixed around its outer edge to the frame either directly with corrugations or pleats, or by means of a flexible roll of cloth, sponge or rubber. The purpose of these is to permit forward and backward motion of the cone while holding it firmly against any sideways movement, but they also have another important function. When the cone vibrates, ripples can spread out from the centre like ripples in a pond when a

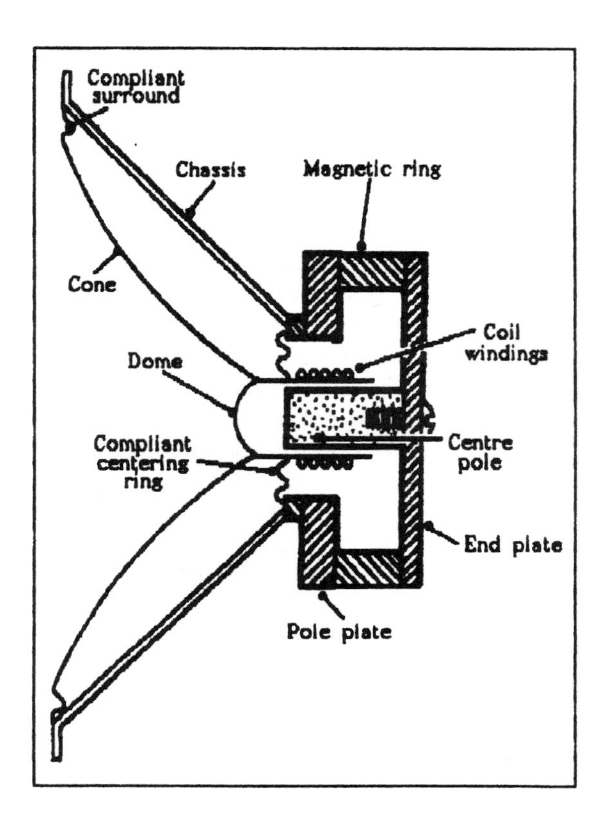

Figure 55 The moving-coil loudspeaker.

stone is thrown in. On reaching the edge they can be reflected back toward the centre like pond ripples rebounding from a side wall

These reflections produce spurious cone motion that is not in response to any electrical output from the amplifier, thereby distorting the sound. Thus a necessary function of the surround is to absorb and dampen such ripples, so preventing reflections.

Cone materials

Ideally, ripples and flexures should be inhibited before they even get to the surround. Hence the cone should be made as stiff as possible; ripples cannot disturb a pond surface that is frozen hard! A perfectly stiff cone would move like a piston, backwards and forwards without any flexures, and so should radiate air pressure waves that are a perfect replica of the electrical currents flowing through the speaker circuit.

This is why cones made of metal such as aluminium have been tried, but they have merely exchanged one set of problems for another. Metal cones tend to 'ring' like a bell when subject to vibration. An ideal loudspeaker cone should have no sound of its own at all; if it has, it will colour the sound it reproduces.

Another problem is inertia. The loudspeaker cone must accelerate and decelerate very quickly in order to produce the very fast vibrations that make up a complex sound wave. To do this its mass must be low. A motor bike will always be away more quickly from the lights than an articulated lorry in spite of having a much smaller engine, because its mass is a tiny fraction of that of the lorry. Metal cones, even aluminium ones, are much heavier than other materials commonly used, so they have a disadvantage here too. Honeycombed aluminium has been tried, being light and about a thousand times more rigid than paper, but has not proved popular.

Cones made of polystyrene reinforced with aluminium foil which is light and rigid, have also been produced. Their snag is poor damping, having a characteristic sound when subject to vibration.

Bexetrene has been used in many hi-fi speakers, being stiffer and more consistent in its characteristics than paper. It too has poor damping and needs to be coated with a plastic damper to tame it. Polypropylene is a more recently employed material which seems to have advantages. It is light, has good self-damping, and is more rigid than paper.

So we come back to paper again. When the paper cone of a loudspeaker is tapped all that is heard is a dull plop without any readily identifiable sound. This is the ideal for uncoloured reproduction. It is also very light, so its lack of stiffness is the only major snag. But we shall see later, this can be made use of and turned into an advantage.

Cone resonance

Every physical object has a fundamental resonant frequency at which vibration is greater than at any other for the same input of energy. Loudspeaker cones are no exception, which means that sound output at the resonant frequency is greater than at all others. The result is an uneven frequency response with an unnatural emphasis at that one frequency.

Below the cone resonant frequency, the sound output falls off at a rate of 6 dB per octave, so the frequency response in the bass region is determined to a considerable extent by the resonant frequency which should therefore be as low as possible for a good bass response.

The resonant frequency in free air is proportional to the square root of the reciprocal of the mass of the cone times the compliance of the suspension. The formula is:

$$f_r = \frac{1}{2\pi\sqrt{(MC)}}$$

in which M is the mass in grams and C is the compliance in metres per newton.

Compliance, which is the opposite, hence the reciprocal, of suspension stiffness, can be calculated from the cone mass and the resonant frequency as follows:

$$C = \frac{1}{(2\pi f_r)^2 M}$$

Thus the compliance (the opposite of stiffness) and the mass should be high, but if they are too high other problems can arise. If the suspension is too compliant it may not keep the cone in place at high volume levels, while if the mass is too great, more energy is required to move the cone, hence the speaker sensitivity is low and a large amplifier power is needed. The resulting high power dissipation in the coil causes heating and possible failure. Furthermore, as a large mass results in high inertia, the cone will not respond to rapid high-frequency electrical signals.

Delayed resonance

In addition to the resonance due to the effect of mass and compliance there is another. When ripples move outward from the centre of the cone to the rim and are not completely absorbed by the suspension, they are reflected back to the centre. When the cone radius equals one wavelength or a multiple of it, the contours of the outward and reflected ripples coincide to produce an apparently stationary ripple or undulation of the cone. This is known as a *standing wave*.

However, when the applied electrical signal ceases, the standing wave subsides, and the consequent cone motion radiates sound as it does so. Stored energy is thus released as spurious sound after the cessation of the signal. The effect is thus termed *delayed resonance*. For an 8 in (20 cm) cone, the fundamental delayed resonance is at 4 kHz with harmonics at 8 kHz and 16 kHz. Efficient absorption by the cone surround is vital to minimize standing waves and the resulting delayed resonance.

Another spurious motion performed by some cones at certain frequencies is what is known as the *bell mode*. With this, opposite quadrants of the cone perform a flapping movement in unison, moving backwards and forwards together while the adjacent quadrants flap in the opposite directions. However, two lines at right angles across the cone which define the boundaries of each quadrant remain stationary relative to the flapping. This effect is due to lack of stiffness of the cone itself. Both these effects are considerably reduced in the elliptical loudspeaker.

At the centre of the cone is a dome that serves as a dust shield to prevent foreign particles from getting into the air gap and causing grating noises. At certain high frequencies, this dome sometimes moves independently of the cone, by reason of the compliance of the glued joint. It thereby exhibits its own resonant frequency, which colours the reproduction. To avoid this in some models, the dome is moulded as an integral part of the cone.

The coil

Under the dome at the apex of the cone lies the coil, which consists of a number of turns of copper wire wound on a paper, composition, or aluminium cylinder.

To reduce the mass and thereby the inertia in high frequency speakers, aluminium is sometimes used instead of copper wire. To get as many turns as possible within the magnetic field, the wire is often of square, hexagonal, or ribbon configuration, so permitting more turns per inch. Up to 40% greater conductor density can thereby be achieved, thus making for a more efficient motor system.

The standard impedance of the coil is 8 Ω, but 4 Ω and 16 Ω models are also available. Formerly, 3 Ω was the standard with 15 Ω for larger units, and these may still be encountered. The impedance consists of resistance and inductance in series, the resistance making up approximately two thirds of the rated impedance. So the impedance of an unknown speaker coil can usually be determined by adding half as much again to the measured d.c. resistance.

The impedance may be considered a minimum value as it rises to a peak at cone resonance, and is usually above the rated value over most of its frequency range. This is of no great importance, but if the impedance should drop below the rated value it could cause overloading of the amplifier. Usually, a higher impedance than that of the amplifier output rating means less power, but often lower distortion, whereas a lower impedance produces higher distortion and possibility of amplifier overload with damage to its output stage.

Cone centring

At the back of the cone, there is a ring of flexible material with corrugations that is secured to the framework at its outer edge and to the cone at its inner. This has the important function of keeping the cone centred relative to the magnet poles.

An off-centre cone produces distortion as it rubs against the magnet pole and can be checked by standing the speaker on its magnet, face upward, and gently pressing the cone inward with the thumbs at opposite points across the diameter, then releasing it. Any rubbing can usually be felt, or heard if an ear is placed close to the cone. Sometimes, though, trouble may be experienced from loose coil windings and these may not be detected by this test.

The magnet

The magnet usually consists of a magnetic ring or rod mounted axially at the back of the speaker. The front pole is terminated by a steel rod pole piece which penetrates inside the coil, and is only slightly of smaller diameter so that the air gap between it and the coil is small. The rear magnet pole is extended by a cylinder or U piece toward the front, where it terminates in a plate with a hole or a ring that surrounds the outside of the coil (Figure 56).

The magnetic field is thus concentrated between the internal rod and the inside of the surrounding hole and thereby through the coil windings, As with all magnets there is a small external leakage field, but modern speakers are designed to reduce this to a negligible amount. This accounts for why the magnets on some loudspeakers seem weak judged by their attraction for external metal objects.

Connections from the coil are taken to a couple of soldered blobs on the cone from which highly flexible stranded copper wires connect to a terminal strip on the speaker frame. These wires must never be tight, nor must they loop down to touch the cone at any other than their soldered connection; they must be completely free of all obstruction. Failure to ensure this can result in buzzing noises as the cone vibrates.

Dedicated drivers

To achieve an extended low-frequency response, the mass of the cone needs to be large so that it has a low resonant frequency. Furthermore, its diameter should be large because the efficiency of the cone falls with decreasing diameter at low frequencies. However, to obtain a good transient and high-frequency response the cone should be small and light.

These conflicting requirements have led to the general use of separate drivers for treble and bass, commonly known as tweeters and woofers. The signal is split into two, one containing all the high frequencies and the other the low by

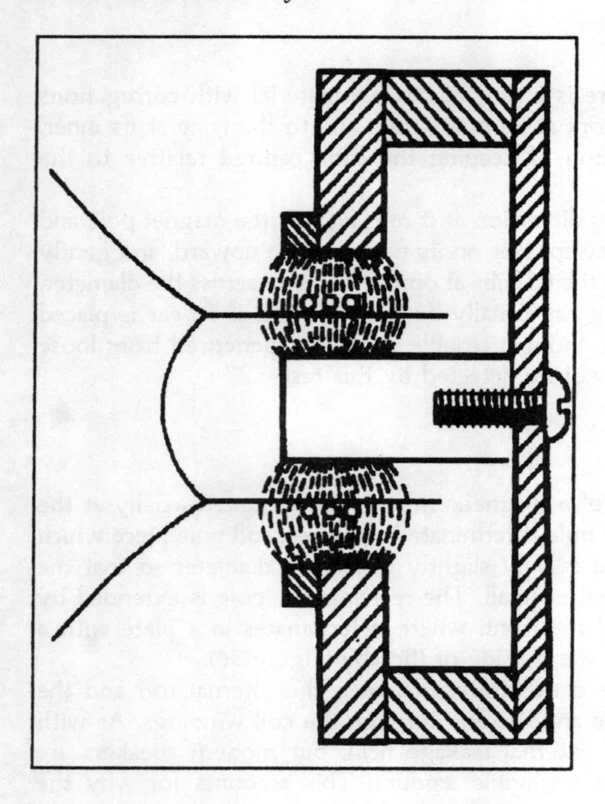

Figure 56 Close-up of coil in magnetic field.

a filter circuit termed a *crossover network,* and fed to the respective drivers. A mid-frequency range speaker is also used in some models, and some have super-tweeters and sub-woofers to extend the range to inaudibility in the treble and bass.

Controlled flexure

There are many disadvantages in using separate bass and treble units. These arise from their spatial difference which causes mutual interference and cancellation effects, distortions caused by the crossover circuit, and difficulties in maintaining balance over the whole frequency and power range. Even for hi-fi purposes, the use of a single unit is to be preferred.

With public address speech reinforcement systems, single loudspeaker units to cover the whole frequency range is almost universal. Yet in theory a single speaker cannot cover a wide range; how then does it do it?

The answer lies in the flexure of the cone at different frequencies. At high frequencies, the central area of the cone responds, but the rest of the cone

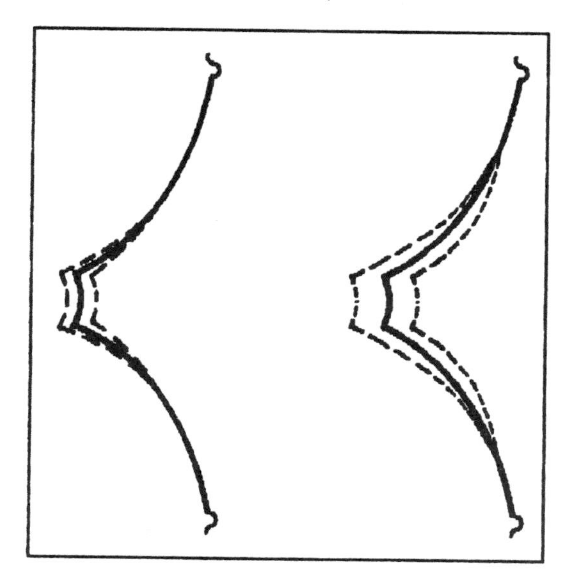

Figure 57 Controlled flexure. Central cone areas move independently at high frequencies. At lower frequencies larger areas come into play.

remains stationary because of its inertia. This independent movement of the central area is possible because of flexure of the non-rigid cone around that area. As the frequency decreases, so larger areas of the cone are brought into play, until at low frequencies the whole cone is in motion (Figure 57).

This effect occurs to some extent with most loudspeakers, but some cones are specially made to exploit it. These have curved sides, and the flexure points are designed into them so that a smooth coverage of a wide frequency range is achieved. They often have a small horn fixed to the centre of the cone to increase efficiency at high frequencies.

Although not having quite the range of separate drivers, it is ample, a typical specification being 40 Hz to 17 kHz. Such full-range drivers avoid all the problems of having multiple drivers, and have few vices. They have thus much to commend them for hi-fi use. For public address, though, ordinary loudspeaker units have a perfectly adequate frequency range, and extended range units, besides being unnecessary, could cause feedback problems.

Cone velocity and radiation resistance

As the frequency rises and the cone makes more excursions per second its speed must increase to maintain the same amplitude. This needs more power, but if the power is constant at all frequencies the speed must also be constant. So when the frequency rises, the amplitude of the cone excursions must decrease to maintain the same speed. This means that the output diminishes as the frequency increases, an effect which without compensation would give a very poor treble reproduction.

Fortuitously, there is another defect in the way sound is propagated by a loudspeaker cone that almost exactly cancels the effect of the first. This is *radiation resistance*. At low frequencies the cone is an inefficient sound radiator. It pushes the air out of the way instead of compressing it into a sound wave.

As the frequency increases, the air does not move aside fast enough to avoid compression but offers a resistance to the cone and so produces sound. The higher the frequency, the greater the radiation resistance and the more efficient the air coupling to the cone. Thus the acoustic output rises and exactly compensates for the diminishing cone excursions due to fixed cone velocity.

The compensating effect works up to a point when the radiation resistance is at a maximum and cannot increase further. This frequency range is termed the *piston region* of operation. Above this the response begins to fall off because the cone excursions due to velocity effect continue to decrease. However, cone flexure effects maintain the response further, and also the beaming effect at high frequencies increasingly concentrate the sound in front of the cone. Thus a useful response continues well above the piston region, so making full-range single-unit speakers viable.

The piston region transition point is dependent on the diameter of the cone. For a flat disc radiator in a true infinite baffle, the relation between the transition frequency and the cone diameter is

$$f = \frac{68\,275}{\pi d}$$

in which *d* is the cone diameter in centimetres. For *d* in inches,

$$f = \frac{26\,880}{\pi d}$$

The baffle

When the cone of a moving-coil speaker moves backwards and forwards it generates two separate sound waves, one at the front and the other at the back. Air is compressed in one direction while that in the other is rarefied. The two waves are thus out of phase.

Cancellation occurs when the front and rear waves meet at the rim of the loudspeaker. Radiated sound then consists only of high frequencies having a wavelength shorter than the radius of the cone, as one or more complete cycles of these are propagated before cancellation occurs at the rim. This accounts for the familiar tinny effect when a loudspeaker is operated without a baffle.

It is evident that some means must be provided to keep the two out of phase waves physically apart, and an obvious way to do so is to mount the speaker on a large flat board termed a baffle. They still meet at the edge, but they have

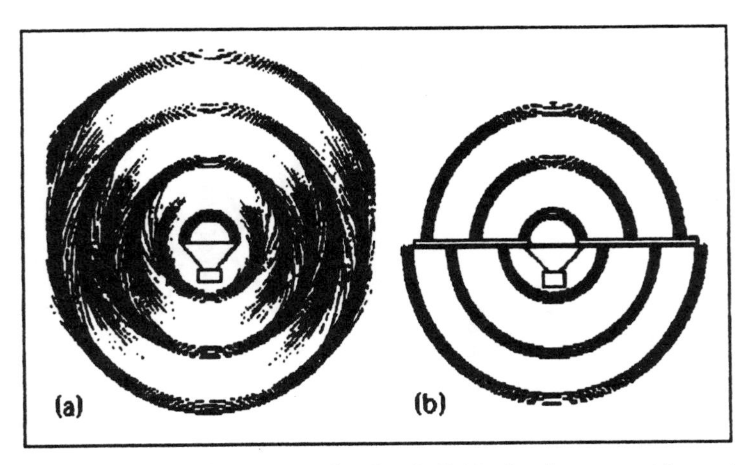

Figure 58 (a) Cancellation of sound with unbaffled loudspeaker; areas of low and high pressure flow into each other. (b) A baffle separates the two waves.

further to go and so longer wavelengths can be propagated before cancellation takes place. Thus the bass response is extended compared to that of an unmounted speaker (Figure 58). The flat baffle has many advantages, among which is the lack of air resonance that produces the coloration inherent with an enclosure. Also panel resonances and vibrations common with cabinets are minimal.

At wavelengths that are half the radius of the baffle, a compression wave from the rear is propagated at the same time as the next compression wave from the front, so that they actually reinforce each other, At whole wavelengths the opposite occurs and cancellation takes place.

Reinforcement thus occurs at wavelengths that are 0.5, 1.5, 2.5, ... times the radius of the baffle, whereas cancellation takes place at whole multiples, 1.0, 2.0, 3.0, ... times the radius. The result is a very uneven frequency response with alternate peaks and troughs throughout its range.

The effect can be easily avoided by simply mounting the speaker off centre on the baffle. There is thus no uniform radius and the cancellation and reinforcement effects are smoothed out. A rectangular baffle with the speaker off centre gives good results, while a circular or square one with the speaker at the centre is the worst possible case (Figure 59).

The big problem with an open baffle lies in the size of baffle needed to procure an adequate bass response. To achieve a flat response down to 45 Hz requires a baffle with the shortest radius of 25 ft (7.7 m). That means a width and height of more than 50 ft (15.4 m) which is obviously impractical.

Looking at more practical dimensions, a 2 ft (0.6 m) radius which is a 4 ft (1.2 m) width or height would be about the maximum. This would start to roll off at 280 Hz, which is rather high, but an ameliorating factor is that the bass fall-off is only 6 dB per octave. This means a -6 dB response at 140 Hz, and

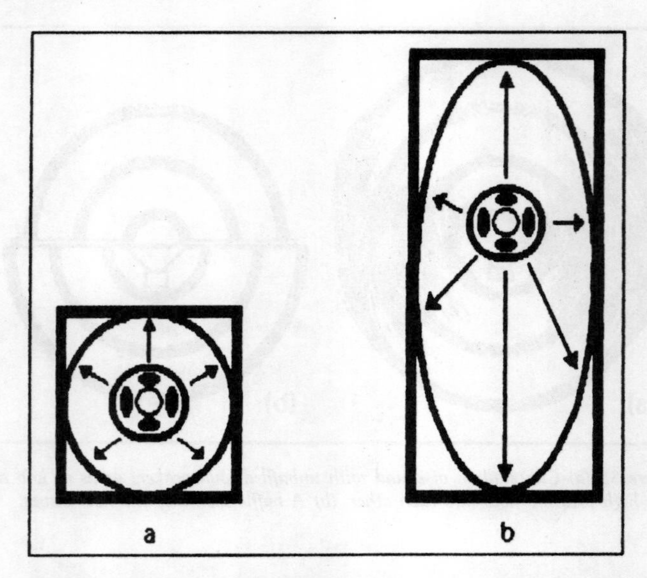

Figure 59 (a) A speaker in the centre of a square baffle has almost equal paths to the edge resulting in reinforcement and cancellation at whole and half multiples of the wavelength (b) On a rectangular baffle the unequal radii smooths out the response.

a — 12 dB level at 70 Hz, so there is at least some response in the bass, though not very much. Fortunately, though, an extended bass response is unnecessary for speech, and in fact is detrimental to clarity and likely to provoke feedback. Most public address operators cut the bass to a certain extent anyway to avoid these effects.

Adding sides

The dimensions of a baffle can be practically increased by adding sides, a top and a bottom. The front-to-back path is thereby extended. A further step is to add a back in which a number of slots have been cut. This adds to the front-to-back path and the slots serve as an acoustic resistance to the rear wave. So the bass is noticeably increased. However, the enclosed space has a more pronounced resonance which affects the reproduction. Alternatively, the back can be solid, thus blocking off the rear wave completely. However, the back itself can resonate, unless it is of substantial thickness and screwed at frequent intervals around its perimeter.

Really, we now have a conventional loudspeaker cabinet such as used for ante rooms, and other auxiliary locations. These come in various shapes and sizes including a V shaped configuration that contains two units angled downward, one on either side. This is intended for corridors or other narrow areas.

From the viewpoint of an effective baffle, ceiling mounting scores highest because the ceiling itself becomes the baffle. It has a very large area with virtual total separation of rear and front waves, an almost true infinite baffle. In theory it should therefore have an extended bass response if there are no air leaks in the mounting.

However, this is not important for public address work and in fact, as we have seen, is not really desirable. It has though the advantage of being free from box resonances, but it can strongly excite the vertical auditorium resonance in the bass region. The overall effect of conventionally mounted ceiling loudspeakers is therefore rather bassy with corresponding loss of clarity.

Horn loudspeakers

The horn loudspeaker was at one time the mainstay of the public address system. Now it is not often seen other than for announcements at sports stadia. It can have other applications, although these are rare. We will take a look at it though, if only to understand why it is not generally used today except for outdoor events.

The horn consists of a small moving coil unit usually with a metal diaphragm or cone feeding into a long tube which gradually increases in diameter until it opens out into a wide flare. The big advantage is its very high efficiency. It is the most efficient means yet devised of coupling the motion of a loudspeaker cone to the air with minimal loss.

Flares

The manner by which the horn increases in area can affect its performance. The simplest configuration is a cone, but it is by no means the best. Reflections can occur between the sides, and these cause interference and irregular frequency response as well as distortion. The ideal is an exponential horn by which the area increases according to an exponential law. This gives optimum air load matching and prevents internal reflections (Figure 60).

The law governing the expansion of area implies that there is a fixed relationship between the length of the horn and the size of the flare at its end. A large flare must have a long passage leading to it.

Another factor related to size is the frequency response. The shortest wavelength that the horn will reproduce is twice the diameter of the throat of the horn. The longest wavelength it will radiate is equal to twice the flare diameter. It is this last fact that gives rise to the big disadvantage. To obtain a response down to 100 Hz, a flare of 5.6 ft (1.7 m) is required. For a 50 Hz response it would have to be 11.2 ft (3.4 m).

The length of the horn needed for a specified flare and throat area is

$$L = \frac{1575\ (\log_{10} A - \log_{10} a)}{f \log \varepsilon}$$

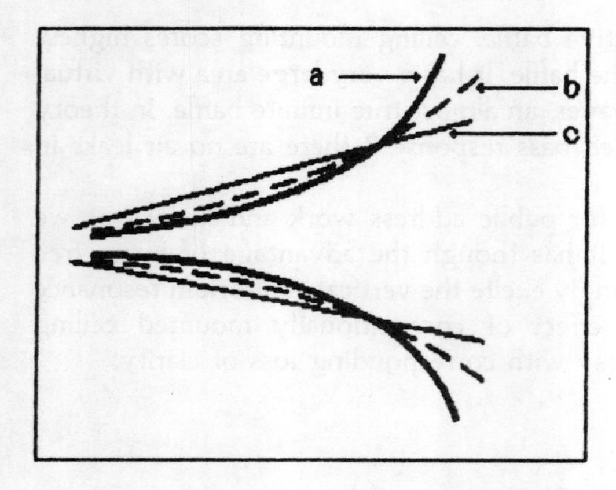

Figure 60 Various horn flares: (a) hyperbolic; (b) exponential; (c) conical.

In which L is the length of the horn in inches; A is the area of the flare, and a the area of the throat (the units are immaterial provided both are the same); f is the lowest frequency; log ε is 0.4343. For L in centimetres, replace 1575 in formula by 4000.

A horn following a hyperbolic area increase gives a response to a lower frequency than that of the exponential horn, but the roll-off below it is more rapid. The area increase from the throat is more gradual, so the sound pressure is greater there, to fall off more rapidly near the flare. This pressure variation along the length results in distortion being generated.

Throat design

The throat needs to be of as small an area as practically possible in order to obtain a good high-frequency response, because as we have already seen the shortest wavelength the horn will produce is twice the throat diameter. However, the cone needs to be larger than this in order to function effectively, so this means that the area immediately in front of the cone must narrow down to the start of the horn proper. A region of high pressure is thereby created in front of the cone which could cause it to respond in a non-linear fashion and so produce distortion. To avoid this the pressure is equalized by a sealed chamber placed behind the cone.

Another problem is that sound pressure from the central and outer areas of the cone could arrive at the centrally located throat at slightly different times because of the difference in spacing from it. Cancellation effects at various frequencies could thereby occur. This is prevented by introducing a plug with holes in it in front of the cone to delay some of the pressure waves so that they all arrive at the throat at the same time.

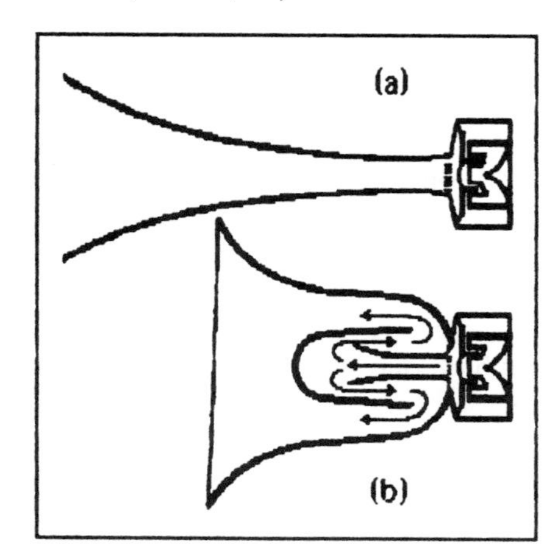

Figure 61 (a) Exponential hord; (b) re-entrant horn.

The re-entrant horn

The big disadvantage of large size to achieve a reasonable bass response can be reduced to some extent by folding up the length. The result is known as the re-entrant horn (Figure 61).

The sound emerges from the throat and is deflected by a metal cone to the back of the flare from which it emerges. The flare can be either circular or rectangular. The former is more acoustically correct, but the latter can be more convenient to handle and mount.

Directional characteristics vary, but 60° to 100° is typical. One model which had a flattened flare and front diffuser was claimed to be omnidirectional. Power handling capabilities range from 5 to 40 W.

While the re-entrant configuration reduces the size of the horn length to manageable proportions it does not help the flare size. As we have seen length and flare diameter are related, so increasing one without the other gives little advantage. Most horns have flares under 2 ft (0.6 m) across, and so have little response below around 300 Hz. While this is sufficient to give intelligible speech, it makes it sound hard and tinny. This is of little consequence for short announcements at outdoor events, but for more prolonged use would be unacceptable,

Efficiency

However, their high efficiency, which can be over 80% (percentage of sound power produced from a given electrical power) compared to just a few per cent of other types of loudspeaker, gives a very high output and so a long range as well as a high degree of audibility over ambient noise. They are thus well suited for announcements at sports stadia. A nest of horns mounted in an elevated position and suitably directed can cover all of the spectator area.

A point to consider though is that a horn can be ear-shattering at close quarters, so one should never be mounted for high-power use close to a spectator area. The power should be used to achieve a large area of coverage by forming a pool of sound from a height.

The efficiency can be used to good effect when power supplies are limited and high output is needed at the sacrifice of quality. Thus small re-entrant horns are used for battery-powered hand-held loud-hailers.

Horns were once used in the cinema to give high-power coverage over the whole auditorium for the relatively small amplifier powers that were then available. Amplifier power is no longer a limiting factor, but it is feasible to use a single horn in this way. The horn would have to have a flare of at least 6 ft (2 m) diameter to give good low-frequency response, and if mounted high above the screen, would cover the whole auditorium with no interference or blind spots.

The line-source loudspeaker

The single-unit loudspeaker on a baffle behaves as a doublet or dipole in that it radiates sound in a figure eight configuration with equal but out-of-phase lobes appearing from the front and the back. If the unit is in a sealed cabinet or is ceiling mounted, the rear lobe is suppressed or reduced, or it is dissipated in the above-ceiling area.

The remaining front lobe projects forward from the loudspeaker as a sphere. Expansion of the pressure wave through three dimensions produces a reduction of amplitude with distance of 6 dB for each doubling of distance. Off-axis sound pressure with the doublet is equal to the cosine of the off-axis angle, times the on-axis pressure at 1 kHz:

$$SPL_2 = SPL_1 \cos \theta$$

where SPL_1 is the on-axis sound pressure level; SPL_2 is the off-axis value, and θ is the angle.

Some practical examples of this are: at 25° the off-axis *SPL* is still 90%; at 41° it is 75%, at 60° it is 50%; while at 75° it is 25%. It can be seen from this that the level decreases only slightly over the first 25° deviation, but falls off increasingly rapidly thereafter. At high frequencies the angles are less, while at low they are greater.

If we arrange a number of loudspeakers in a vertical column the propagation pattern changes from a doublet to that of a line source. Sound is concentrated in the area immediately in front of the cones and there is little radiation above and below. Instead of an expanding sphere, we now have an expanding cylinder of sound. Being thus confined to two dimensions instead of three, the pressure

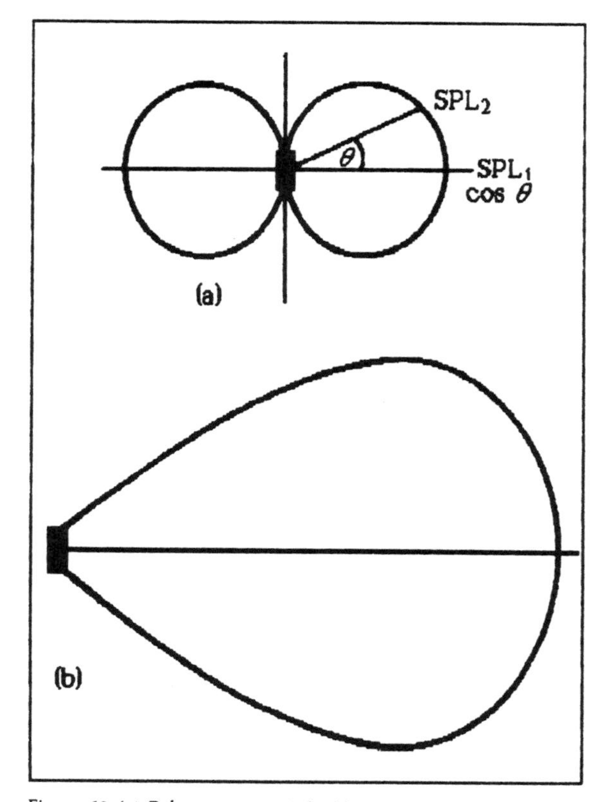

Figure 62 (a) Polar response of doublet. Off-axis SPL is equal to on-axis x cos A. (b) Polar response of line-source.

expends over a smaller area and so projects further in the horizontal plane than a single unit. It actually radiates twice as far for the same *SPL* reduction as the doublet.

So while the sound pressure from a doublet drops 6 dB for each doubling of distance, the line source drops only 3 dB. The level falls by 6 dB for four times the distance. This is a very useful factor and it means that a column can cover a much larger area than a doublet or single loudspeaker. It is especially useful for large auditoria in which it would be otherwise difficult to provide sufficient coverage for the central areas. It also means fewer loudspeakers, so saving installation costs.

While the vertical dispersion is concentrated into a narrow beam, thereby increasing the forward coverage, the side or horizontal dispersion is much the same as that of a doublet, and the SPL is approximately proportional to the cosine of the off-axis angle. It actually is flattened a little at the sides and the end is drawn out to give more of an oval configuration (Figure 62 and 63). The side response thus falls off more rapidly after the 25° and this must be taken

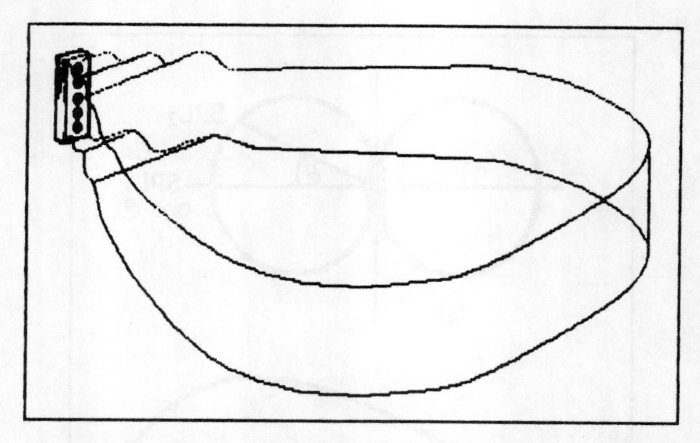

Figure 63 Sound distribution from column speaker showing lobes top and bottom.

into account when siting column speakers to ensure that off-axis areas are sufficiently covered.

Divergence and frequency

The divergent angle of the beam at its top and bottom depends on the length of the line source, that is the distance between the two end drivers, and also the wavelength of the sound. A reinforcement and cancellation pattern occurs between the pressure waves produced by each unit in the column. Maximum reinforcement occurs along a line produced directly from the centre of the column, whatever the wavelength, but cancellation effects start on deviating from that on-axis line, and increase as the angle from the line widens. Cancellation is virtually total at about 58° when the column length equals the wavelength, so this is the effective maximum dispersion angle at that particular frequency. When column length is twice the wavelength, the angle at which total cancellation occurs is 29°, which is thus the maximum dispersion angle, when it is four times, the angle becomes 14.5°, and so on.

Divergence thus narrows for a particular column length as the frequency rises, until it decreases to form a parallel beam at the highest frequencies. To take a 5 ft (1.6 m) column as an example, the frequency corresponding to this wavelength is 220 Hz, so the column has a 58° dispersion angle at that frequency. At 440 Hz, the angle is 29°, at 880 Hz, it is 14.5°, and at 1 760 Hz it is 7.25°.

Measurement of the divergence angle using mixed frequencies such as pink noise, gives a −6 dB level at around 14° for a 5 ft column. This is about the angle calculated for 1 kHz, which is the halfway point of the 100 Hz to 12 kHz range of a public address column. So the average divergence angle can be taken as 14° for a 5 ft column. It is narrower for a longer one (Figure 64).

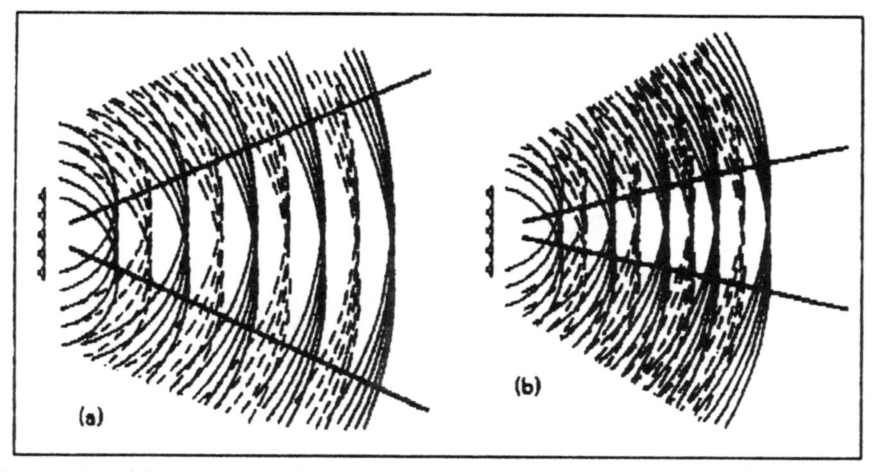

Figure 64 Sound dispersion from column speaker. (a) Cancellation of high- and low-pressure waves (solid and dotted lines) produce an angle of 58° when wavelength is equal to column length. (b) It is 29° at half that wavelength. So the beam narrows as frequency increases.

A beam that is only slightly divergent is a useful feature of the column speaker because it can be directed where the sound is required, into the audience; little reaches the walls and ceiling, thereby giving fewer reflections. This gives better intelligibility and less feedback than single units. However, care must be taken in their siting and angling.

Tapering

In some installations using long columns, where optimum angling is not possible, the narrow beam at high frequencies could produce poor intelligibility in areas not directly in line.

To obtain wider high-frequency vertical divergence, columns can be frequency tapered. This means that only the centre drive units are fed with the full high-frequency range, while adjacent ones are fed progressively less until the outer ones are reproducing low frequencies only. Thus the effective column length is progressively shortened as the frequency rises.

The means of achieving frequency tapering depends on the method of connecting the drive units. If connected in series, capacitors of increasing values can be connected across the outer loudspeakers with the largest value across the final ones. If connected in parallel, inductors of increasing values can be connected in series with the outer units with the largest value at the ends (Figure 65). Longer columns having many units may have a series–parallel arrangement, and with these a combination of capacitors and inductors may be used.

Alternatively, a short line of tweeters can be mounted at the centre of the column alongside the main drivers, these being fed by a capacitor, or a more complex cross-over unit. Some commercial units use this method.

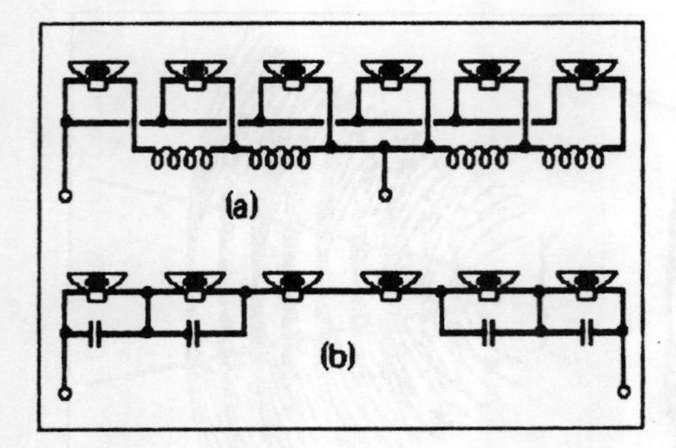

Figure 65 Frequency tapering: (a) with parallel drivers; (b) with series drivers.

Small lobes usually appear at the start of the divergent beam at the top and bottom which are due to the off-axis radiation of the end units. These are sometimes suppressed by power tapering the column. This consists of feeding the maximum power to the central units, with a gradually reducing amount proceeding outward, ending with minimum power at the ends.

One method of doing this in a six-unit column is to connect all the units in parallel, with the input going directly to the centre pair; an inductor is connected from these to each of the next pair, and a further inductor from those to the end ones. The column has thus a very low impedance and must be supplied via a suitable matching transformer.

The inductors also give frequency tapering, so controlling the total vertical polar response by eliminating the lobes and giving a divergence angle that is more or less the same for both high and low frequencies.

With the more usual series connected column, power tapering can be achieved by shunting resistors across the final and penultimate drivers, a value equal to the driver impedance across the last ones, and one that is double the impedance across those previous. Some amplifier power is thereby wasted though.

Frequency and power tapering, while tidying up the technical polar response, confers little real advantage. Neither non-divergent high frequencies nor lobes pose any real problems if the columns are properly positioned. Most commercial units have neither.

Construction

When constructing a column there are no special rules to observe as regards number of units or spacing, except that the spacing should be equal. Elliptical units are very suitable as they give length with little width, and so when mounted vertically give the closest approach to a true line source. The power is divided between them, so six units each having a power rating of 3 W will produce a column capable of handling 18 W. This is more than adequate for most applications.

If the drivers are the standard 8 Ω impedance, series connection will give a total of 48 Ω which may seem high; however, it is common to use four columns in medium-sized halls (those seating about 200), so these connected in parallel would give a combined impedance of 12 Ω, which just loads a low-impedance amplifier nicely from the 4 Ω tap and gives spare capacity for a hearing-aid loop.

For use on a 100 V line, the impedance should be matched to that of the transformer winding. Although error does not result in a mismatch or power loss, it produces inaccurate wattage tappings. Line transformers have secondary impedances of 4, 8, and 16Ω, though many do not have the 4 Ω tap. These values can be matched by loudspeaker drivers only in multiples of four. Four 8 Ω units in series parallel give a total of 8 Ω, while eight give 16 Ω.

Six units give either 12 Ω or 5.3 Ω depending whether they are connected as two parallel groups of three in series, or three parallel groups of two, An error is thus unavoidable with standard transformers.

When wiring the drivers inside a column, care must be taken to ensure that they are all connected in phase. Failure to do so will have some cones moving in while others are moving out, and cancellation will occur within the column. Series units are connected with positives going to negatives of adjoining ones, while parallel groups are connected with like terminals together.

After connecting all the units in the column, double check that the terminals are connected correctly; it is very easy to make a mistake, especially when wiring up a number of columns. The inside should be filled with an absorbent material such as BAF (bonded acetate fibre) to dampen internal resonances.

Phasing

Loudspeaker drive units will be found to carry a mark, either a red spot or a plus sign, to signify the positive terminal. There is of course no positive d.c. voltage of current involved, the term is used to identify the phase of the loudspeaker. When a positive potential is applied to the terminal so marked, the cone moves forward.

If the loudspeakers have no markings, polarity can be checked by using a dry battery. A $4\frac{1}{2}$ V or 6 V lantern battery is quite suitable. Connect one pole of the battery to one terminal of the loudspeaker, and the other pole to a short length of wire. Touch the wire momentarily on to the other loudspeaker terminal and observe which way the cone moves. If the observed movement is too small to detect direction, place a finger lightly on the centre of the cone. It should then be possible to determine whether the cone moves in or out when the wire is contacted. Do not leave the battery connected for longer than necessary to establish direction of cone movement.

If the cone moved outward, the positive pole of the battery was connected to the positive terminal and it should be marked accordingly. If it moved inward, then the battery negative was connected to the driver positive (Figure 66).

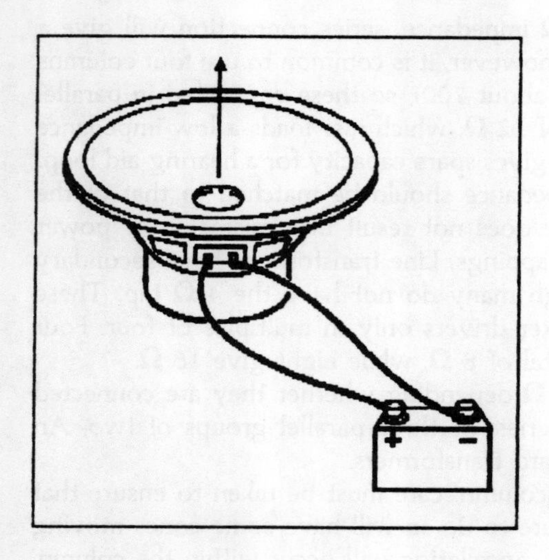

Figure 66 Phasing a loudspeaker. When the positive of the speaker is connected to that of the battery the cone moves out.

9 Public address loudspeaker systems

There are several loudspeaker systems whereby an indoor hall or auditorium can be served with sound from a public-address system. The aim is to achieve even coverage, ample volume, intelligibility, and natural effect as discussed in Chapter 1. The loudspeaker system is the most important factor in achieving this. Some systems are much better than others for certain applications.

Ceiling matrix

Multiple loudspeaker systems once were the standard because there was no practical alternative. They usually took the form of a number of cabinet units mounted along the walls or suspended from the ceiling. When the acoustically superior line-source columns were introduced they fell into disuse for auditoria and were used for foyers, dining rooms and the like for background music and announcements. Now they have reappeared in halls in some areas in the form of a matrix of rows of ceiling-mounted loudspeakers.

When so used, listeners that happen to be exactly equidistant between loudspeakers hear the sound more or less as reproduced by a single one. For those in all other locations, the sound from various nearby loudspeakers is delayed in proportion to the distances travelled. The out-of-phase sounds thus mutually interfere, and produce a sequence of cancellations and reinforcements that chop up the audio spectrum in a comb filter effect.

The ear is not aware of hearing different sources or of the delays because it locks on to the sound that arrives first, providing it is not more that 10 dB lower than one arriving later. So all the sound appears to come from the nearest loudspeaker, and the listener is not conscious of sound from any of the others. This is known as the Haas effect.

The listener may thereby be deceived into thinking that he is actually receiving sound from only one source and that the others are having no effect. Yet this is not so, as the sound quality is impaired by the interference effects.

Low frequencies are comparatively unaffected. The difference in sound paths from two loudspeakers to a listener would have to be many feet to produce appreciable out-of-phase interference at long wavelengths. At such a distance the volume level from the furthest one would be too low to interfere appreciably.

It is when the sound path difference is only a matter of inches, and the sound levels are comparable, that significant interference occurs. For example, a sound path difference of 6.75 in (17.1 cm) is equal to the half wavelength of 1 kHz, so at any location with such a difference there is cancellation of 1 kHz, and at 3 kHz, 5 kHz, 7 kHz and so on. At 2 kHz, there is reinforcement, also at 4 kHz, 6 kHz, 8 kHz upward. Below 1 kHz the sound level falls gradually from the bass to the first cancellation point at 1 kHz.

This comb filter effect of peaks and troughs depends on the path length difference between two nearby loudspeakers and so varies from one seat in the auditorium to another. If the seat location is within a quadrant of four loudspeakers, the response is more complex, being the resultant of all four pressure waves.

The frequency band most affected is that which conveys the speech consonants, thus impairing intelligibility. Those with hearing deficiencies suffer the greatest impairment. As everyone above early middle-age is affected by presbycusis, a large part of an audience may thereby have some difficulty in understanding speech propagated from a ceiling matrix.

Another factor with this type of system is the auditorium resonance. Every auditorium has three frequencies which are the half wavelengths corresponding to height, length and width. These and their harmonics produce peaks in the frequency response.

With most medium to large halls, the length and width are too great to cause any problems. For example, a 50 ft (15m) dimension has a resonant frequency of 11 Hz, its second harmonic, 22 Hz, the third, 33 Hz, fourth 44 Hz and so on. A 25 ft dimension has frequencies of twice those values.

These are well below speech frequencies and so do not affect speech reproduction, but the smaller height dimension, can. A height of 9 ft (2.74 m) has a fundamental resonance of 62 Hz The second, third, and fourth harmonics are 124 Hz, 186 Hz, and 248 Hz. This is among the lower speech frequencies and has the effect of emphasizing these at the expense of the higher ones. The result is to give a bassy, boomy effect which impairs clarity and intelligibility.

While resonances can be excited by corresponding frequencies generated in any plane within the resonant area, they are most strongly excited when the propagation is along the particular dimension. The vertical resonance is thus strongly activated by downward-facing ceiling loudspeakers.

This resonance is not damped by the audience or carpeting on the floor as might be thought. All resonant bodies including air between two surfaces, have regions of minimum and maximum particle vibration called nodes and antinodes. To suppress resonance, damping must be placed at the antinodes. where there is maximum movement. It has no effect at the nodes as there is no vibration there to stop.

In the case of a room or auditorium resonance, the node of the fundamental is half way along the dimension, while the antinodes are at the ends. Thus, the audience and the carpet do indeed dampen one antinode of the fundamental of the vertical resonance. But the fundamental at around 60 Hz is not the problem. The second harmonic has its main antinode at the centre with two nodes at the quarter and three-quarter positions. The third harmonic has nodes at the sixth, half and five-sixth positions with antinodes at the third and two-third positions.

Thus the important second and third harmonic antinodes are at the half, and the third and two-third positions between ceiling and floor, and so remain undamped by audience or floor covering.

Added to all this, a further disadvantage of the ceiling matrix is that the row of loudspeakers nearest the platform can direct sound directly at the microphone and thereby encourage early feedback unless placed well forward.

Finally, the desirable natural quality which was our third requirement for a public address system, is totally absent. The sound of the speaker's voice comes from above for everyone in the audience, and in many cases over the shoulder from behind. So there is a conflict between the spatial information fed to the brain from the eyes, and that from the ears. Though audiences can appear to 'get used to' this, it does subconsciously produce mental fatigue and lack of concentration and reduction of attention span.

When seated mid-way between two loudspeakers, a slight turn of the head can bring it closer to one than another, then another turn reverses the effect. The result can be an apparent jump in the sound source between the two loudspeakers due to Haas effect.

Summarizing the multiple speaker system as exemplified in the ceiling matrix arrangement, it produces interference patterns giving comb effect response at mid and high frequencies, so obscuring the most important speech sounds, and excites a strong vertical resonance at the lower speech frequencies, both of which reduce clarity and intelligibility. It also lowers the feedback point by radiating direct sound back at the microphone, and is quite unnatural in its sound source location and effect.

It is thus quite unsuitable for serious speech reinforcement in halls and auditoria, but it does have many other applications. In factories and noisy machine shops a loudspeaker suspended over each working point is about the only means of ensuring that announcements or alarms are heard. The proximity of high-noise sources is likely to mask anything propagated from a distance. For quiet background music and announcements in supermarkets the ceiling matrix is unobtrusive and gives coverage over as large an area as the sound is needed. Also individual units would not be masked by displays of goods. Canteens and restaurants likewise can be effectively served by ceiling loudspeakers.

Single ceiling units in theatre foyers and bars can be used to warn of the re-commencement of the programme after the interval, and also in rest rooms and mothers' rooms. They are quite suited for paging in hotel foyers and common lounges.

Column loudspeakers

The column loudspeaker utilizes the line-source principle with a suppressed rear lobe. The polar response is flat at the top and bottom, so giving a beam with a narrow vertical dispersion. Horizontally, the propagation is similar to that of the single loudspeaker, that is it falls off-axis in proportion to the cosine of the angle. However, as the range is greater, the response is elliptical rather than circular (see Figure 63), page 128.

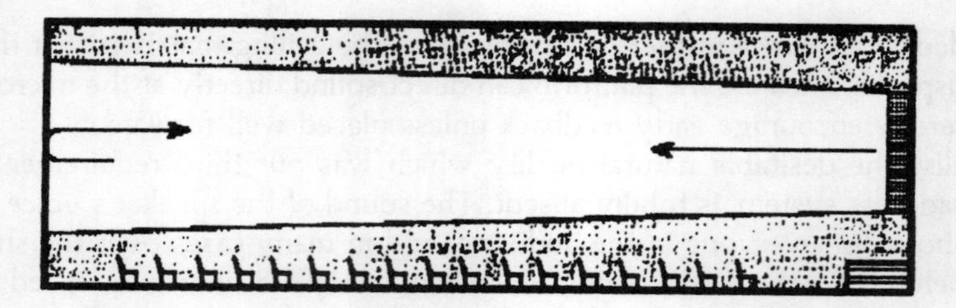

Figure 67 Upright columns too high beam sound over the heads of the audience to be reflected back to the microphone from the rear wall.

Attenuation of sound pressure with distance is half that of a single loudspeaker unit, being 3 dB for a doubling of distance. For single units it is 6 dB.

These features can be used to good effect in a public-address system, offering many advantages over other arrangements. However, the mere use of columns does not guarantee success, they must be used and positioned with due regard to their characteristics. Installations are often seen where this is not the case and result are poor.

One common error is to mount the columns too high with little or no downward inclination (Figure 67). It is evident that the sound is beamed over the heads of the audience, to hit the back wall and so give a strong reflection. The result is an indistinct reverberant sound with high feedback.

If titled at the right angle, all the sound is directed into the audience (Figure 68). All get a strong direct sound and little if any is reflected from the back wall. There is thus minimum reverberation and feedback. The tilt must not be too great otherwise those in the back rows are missed. It can be determined by means of a flat stick temporarily fixed to the top of the column and projecting forward. When seated in the back row, the stick should be pointed directly at the observer with neither its top nor under surface being visible.

Large columns include a larger section of audience for a given angle than short ones. In Figure 68 a column half the size of the one illustrated would not reach to the back row. To do so it would have to be lowered and its tilt angle reduced.

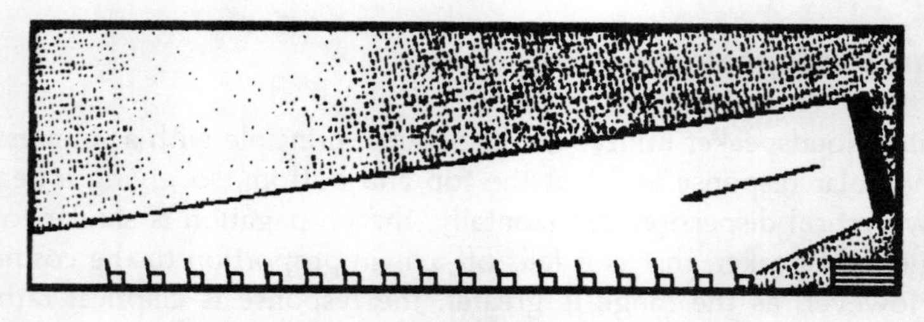

Figure 68 Tilt must be adjusted to take in whole area.

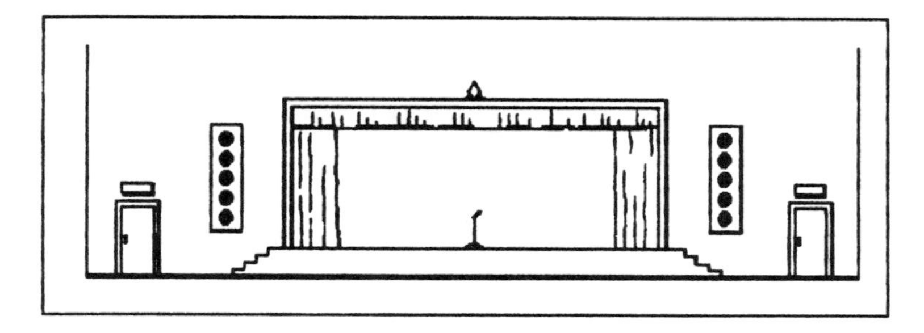

Figure 69 Columns should be mounted low enough to cover a seated audience. They are often too high.

Short column loudspeakers can be mounted upright with no tilt, in which case their bottom unit must be no higher than audience shoulder height. There is likely to be some reflection from the back wall in such a case.

Mounting height, column length, tilt angle and maximum audience distance from the column are thus all inter-related factors. If the vertical propagation pattern is visualized as a flat-topped and flat-bottomed beam as shown in Figures 68 and 63, getting it right should be no problem. Plotting the coverage with scale plans can be a big help.

Centre-fill

Column loudspeakers are usually mounted on either side of the platform or stage (Figure 69). This means that those sections of the audience at the sides are on-axis to one loudspeaker and receive maximum sound pressure from it. The pressure falls off on either side at an angle of 60° by 6 dB. At a point where the 60° lines from each column meet at the centre, reinforcement occurs to give the maximum level; from there toward the back the level is fairly even. This leaves a triangle immediately in front of the stage in which sound pressure is lower, the lowest point being the centre seat of the front row (Figure 70).

In a small hall this is not undesirable, as this area then receives mainly direct sound from the platform with some reinforcement from the loudspeakers at the sides. This gives a very natural effect.

With large auditoriums, direct sound may be insufficient, especially at two or three rows back. Angling the columns serves to reduce the triangle but there is a limit to how much they can be so angled. For wide stages, a single filler loudspeaker may have to be mounted at the front centre of the stage facing the audience.

Extra columns

With large halls, extra columns may be needed further down the hall, otherwise the sound may be rather thin toward the rear. If these are used, the tilt of the

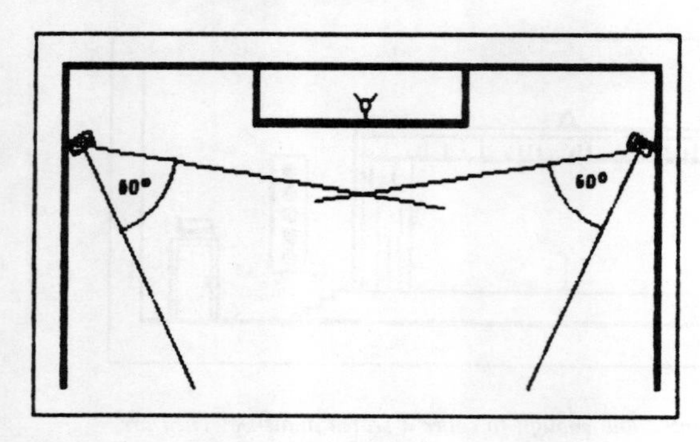

Figure 70 6 dB limit from the normal is 60° so columns should be angled inward to cover the centre.

front columns need only give coverage to where the second pair take over, or a little beyond.

The extra pair can often be driven harder without feedback than the front ones, especially in larger halls. Transformer tappings can therefore be adjusted by trial and error so that each pair has about an equal feedback point. It may be found that when the front pair are just below feedback, the second pair can be tapped up to a higher setting before they initiate feedback.

In mounting them, they should be angled inward to a greater degree than the front ones. Aiming them to face the opposite rear corner is a good rule of thumb, although individual conditions may modify this. Good centre coverage will thus be ensured.

While the individual layout will determine the best place for them, a position about one third of the hall's length from the first pair is usually about right. Spacings of over 50 ft (15 m) between loudspeakers feeding the same area should be avoided as the resulting 45 ms delay from the furthest can be detected as an echo.

Delay lines

If larger spacings are unavoidable, a delay line and slave amplifier can be used to delay the sound to the extra loudspeakers by the same amount as the acoustic delay, so that sound from both arrives at the same time. If the electronic delay is actually made greater, sound from the front columns will arrive at the rear section of the audience before the delayed signal from the loudspeakers although they are nearer.

Haas effect will thereby make the sound appear to come from the columns at the front providing there is no more than 10 dB difference in level. The audience toward the rear will thus be unaware of the presence of the extra loudspeakers at the sides, and a natural front-located source will be achieved for the whole auditorium.

If a delay line is used in this manner, it may be better to reduce the level of the extra loudspeakers instead of increasing it, to ensure that the 10 dB difference, which can override the Haas effect, is not exceeded.

In reverberant surroundings such as railway station platforms, cathedrals and the like, intelligibility is difficult enough to achieve without the added confusion of acoustically delayed signals from spaced loudspeakers. In these situations electronic delay is virtually essential to compensate.

Table 7 Distances and delays at 60°F (15.5°C)

Distance (ft)	Distance (m)	Delay (ms)	Distance (ft)	Distance (m)	Delay (ms)
50	15.4	45	80	24.6	72
60	18.5	54	90	27.7	81
70	21.5	63	100	30.8	90

Line source ceiling array (LISCA)

For an indoor auditorium the choice of loudspeaker system has so far been between the ceiling matrix and the line source. The ceiling matrix we have shown to have many disadvantages, in particular, interference and cancellation effects which play havoc with the frequency response in the all-important upper speech frequency region; excitation of the floor-to-ceiling resonance giving a bassy effect; unnatural sound location; and proneness to feedback.

The conventional line-source system using vertical columns avoids those problems and has been the preferred choice of sound engineers since their inception, but does not give an entirely natural source location. The LISCA system gives advantages even over the conventional column loudspeakers.

Normally, a column should never be used horizontally for two reasons. Firstly, the sharp cut-off at the ends means that coverage is restricted to a narrow band in front of it having a width equal to the column length. Secondly, the wide side dispersion becomes vertical and so sound is directed upwards to the ceiling and upper walls where it produces unwanted reflections.

LISCA consists of a horizontal line source mounted in the ceiling, but it extends the whole width of the hall so there is no problem from end cut-off. The whole audience area is within the width of its beam. The end cut-off actually minimizes reflections from the side walls, and so is an advantage.

It is mounted a little forward from the first row of seats and just off the front edge of the platform. The loudspeakers are not facing downward as with the ceiling matrix, but are tilted to an angle of 64° facing toward the audience.

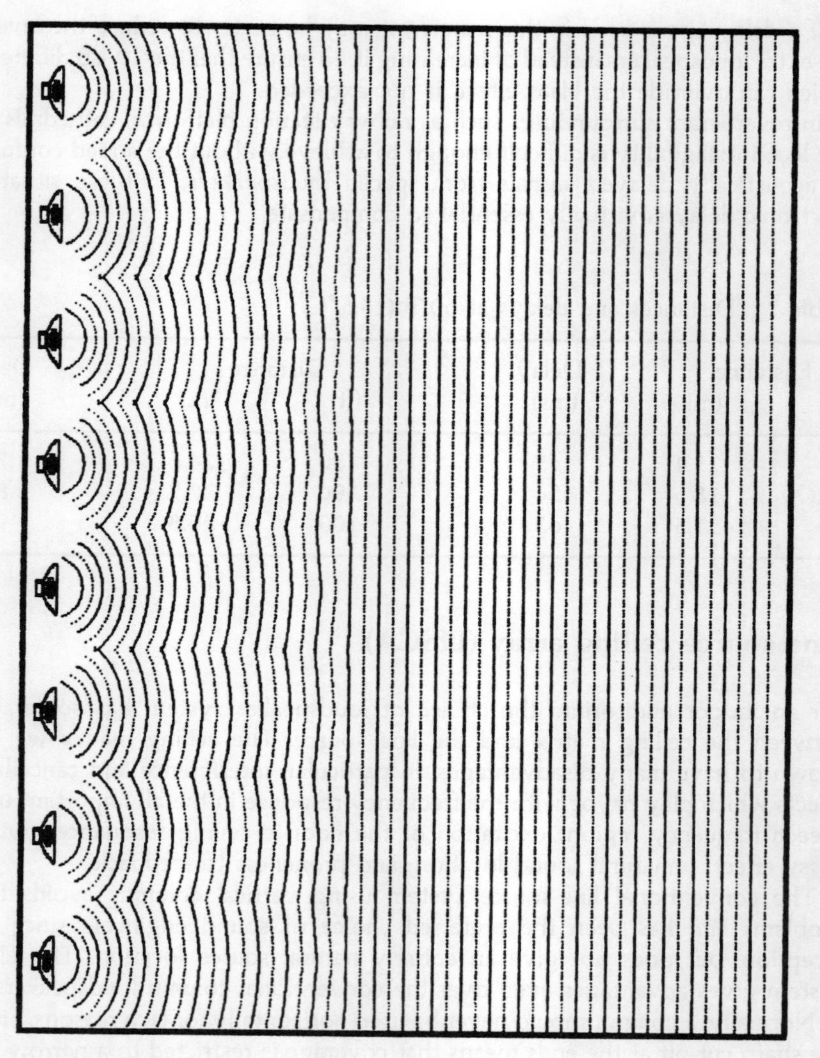

Figure 71 Individual spherical wavefronts merge to form a plane wave a short distance from the array.

A sound pressure wave from a long line source begins as a series of spherical zones, but they quickly merge to form a single flat-fronted plane wave. Instead of circular waves expanding out from several sources, the effect is more like a sea wave coming in on a flat sandy beach. It sweeps along the length of the auditorium (Figure 71).

There can thus be no interference effects, the sound is coherent like a laser beam, so the intelligibility is accordingly high, and is consistent, being exactly the same at all points in the hall. Though mounted in the ceiling the angle of propagation avoids strongly exciting the vertical floor-to-ceiling resonance. The overall result is speech of exceptional clarity everywhere.

Feedback

The tilt of the loudspeakers puts them sideways-on to the platform and the microphone. The angle is thus 90° and the output is cos 90° which is 0. Thus no direct sound at all can reach the microphone from them except at low frequencies. This is in contrast to the ordinary ceiling matrix which, by facing downward, radiates sound from its first row back to the microphone.

As with all systems, there can be feedback from reflected sound from the auditorium, but even this is at a minimum as there is little or no sound directed at the side or back walls. All is directed into the audience where it is required, and most is thereby absorbed.

Sound levels and range

The significant dimension affecting all calculations, is h, the height of the array above the seated audience head level, which is about $3\frac{1}{2}$ ft (1 m). Thus h is the floor-to-ceiling height H, minus $3\frac{1}{2}$ ft. The line source is at an angle of 64° off-axis from the first row of seats beneath it. So the theoretical sound level at that row is $\cos 64^\circ = 0.44$ times that of the same distance along the on-axis line, or -7 dB.

The on–axis line converges with the audience head level at a distance from the array of: $h/\sin(90°–64°) = h/0.44$ or $2.3\ h$. As the propagation loss from a line source is 3 dB for a doubling of distance, the loss along the on-axis line to the point where it reaches the audience is about 3.5 dB.

So with a level of -7 dB under the array at the first row, and -3.5 dB at the on-axis point, the latter is 3.5 dB higher than the former. Using the first row as a reference level of 0 dB, the on–axis point is therefore $+3.5$ dB but may be less according to conditions. Thus the first rows receive less volume than elsewhere, being just enough to reinforce the speaker's natural voice.

As the distance from a listener in the first row to the speaker on the platform is likely to be less than that to the overhead array, the direct sound arrives first and Haas effect makes all the sound appear to come from the platform with little if any awareness of the overhead source. The effect is thus completely natural.

The floor distance from a point under the array to the on–axis point is $h/\tan(90°–64°) = h/0.49$, or approximately $2h$. So $2h$ is the floor distance to the point of maximum sound level.

Beyond this it declines as the line of propagation goes off-axis again and the distance increases. However, distance has less effect than may be expected. SPL decreases with distance from a line source because of the cylindrical expansion of the wave. With LISCA, the wave expands cylindrically from the array until it fills the floor-to-ceiling space. But beyond this no further expansion is possible because it is constrained by those boundaries. The effect is like sound travelling along a tube, and in theory there should be little further loss beyond this point.

There are, however losses due to absorption by the audience, carpet, if fitted, curtains and padded seating. The range limit beyond $2h$ will thus depend greatly on the furnishings. At $4h$ the propagation angle is narrowed to the point where the recessed loudspeaker cones begin to be masked by the ceiling. Low and middle frequencies are diffracted around the obstruction, but high frequencies are not, so intelligibility may start to deteriorate beyond this point. Also, the off-axis angle increasingly reduces the level. So $4h$ can be considered the maximum range to give adequate sound level with highest intelligibility.

If the length of the hall from first row to last is greater than $4h$ or if low ceilings give a low value for h, a second line source is necessary. This does not interfere with the first one if correctly located, and so there is no sacrifice of clarity and intelligibility.

The second line source

The second line source should be located, not where the output from the first is tailing off at $4h$, but where it is strongest at around $2h$ or a little beyond at $2\frac{1}{2}h$. Here, the output from the second array, which is immediately overhead, is -3.5 dB compared to that from the first. So, with most of the sound coming from the first, the perceived location is still forward and there is virtually no awareness of the second array to listeners beneath it.

Moving back from here, the output from the first line source starts to diminish while that from the second increases, thus maintaining even sound level throughout. Furthermore, as the second array is then forward relative to the listener, the natural frontal source location is maintained.

In very long halls or those with a low h, the distance from the proposed second line source location to the last row seats may exceed $4h$. In this case, the location of the second array may have to be modified to $2\frac{1}{2}h$–$3h$.

In this case there may be some awareness of an overhead source immediately below it, as the output from the first is lower than $+3.5$ dB at this point. However, the effect is local, the intelligibility and sound level is not affected. The situation in which the length of a hall from the first row of seats to the last exceeds $6h$ would be unusual.

Having two arrays does not create the interference comb filter effect inherent with the ceiling matrix. Two conditions are needed to produce this in the speech frequency spectrum. First, the volume levels need to be similar, and second, the sound path difference between each source and the listener must be inches rather than feet. While these conditions are common with a ceiling matrix, they are nowhere found with a twin LISCA system. Where sound level differences are small, the path difference is large, and vice versa.

Pseudo stereo effect

An unexpected stereo effect has been noticed with LISCA installations. Not only does the sound come naturally from the front as it should, but it seems to correspond to the location of speaker on the platform.

The explanation seems to be that our auditory system identifies the location of all sound we hear by comparing the phase of sounds received by the two ears. Those arriving first at the ear on one side are perceived to come from that direction; the amount of phase difference identifies the precise angle. This, incidentally, is why it is difficult to locate low-frequency sounds; the wavelength is so long that there is virtually no phase difference from one ear to the other.

Nearby sounds are easy to locate because of their spherical wave front, the curve ensuring that the sound to one ear is well in advance of the other. With more distant sounds, the curvature is much less, and so location is less precise.

With LISCA, the wave front is plane, so there is no curvature; it arrives at both ears at precisely the same time. Therefore there is no lateral directional information. The brain then seeks to replace this missing information with data from another sensor, the eyes. So, when observing participants at any position on the platform, the brain signals its auditory section that that is where the source is.

It is an illusion, but then so is stereo. Nonetheless, it results in an unexpected bonus of naturalness, which together with the desirable graded sound level from the front, the even coverage over the rest of the auditorium, the extremely high degree of intelligibility and clarity, and the low feedback characteristic, makes LISCA difficult to beat for auditorium public address loudspeaker systems. No other system has so many advantages with no drawbacks.

Practical considerations

A line source does not have to consist of a column loudspeaker, that is a box with a row of loudspeakers fitted inside. It can be made up of a row of individual units mounted separately. For LISCA this is usually easier to implement, it requires less disturbance to the existing ceiling, and uses cheaper materials and less of them.

Most modern halls have suspended tile ceilings and LISCA can be installed in these by mounting one loudspeaker in the centre of each square tile across the width of the hall, except the ones adjacent to the wall on each side. Two are also omitted near the centre, usually one either side of the central one or pair, to avoid a high SPL at the centre and give even levels across the width of the hall. Each loudspeaker must be mounted in a box supported on top of the tile, which inclines the unit at the required angle of 64°. Expanded metal and fabric or perforated plastic sheet cover the aperture in the tile so that the array is hardly noticeable. The arrangement of a twin array with SPL levels is shown in Figure 72.

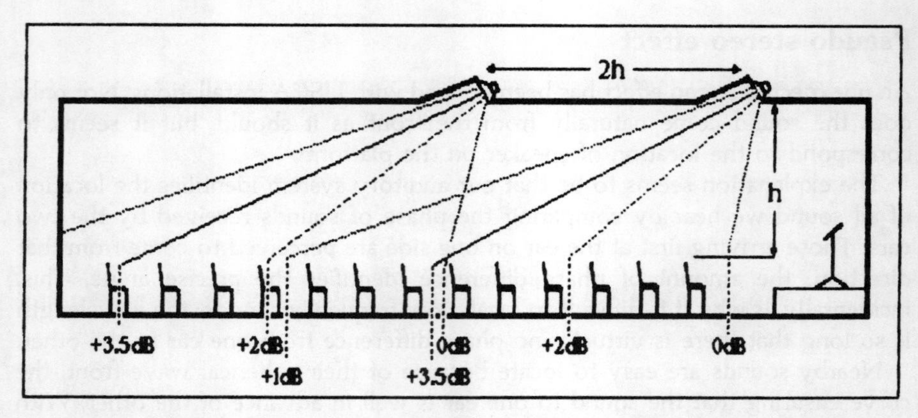

Figure 72 LISCA theoretical sound levels. Increasing from the front, they then remain within 1 dB over the whole area.

BS 6259

This *British Standard for Sound Distribution System,* states: 'It is desirable that the sound should come from one virtual source so that loudspeakers should be grouped.' (11.5.2). LISCA fully conforms to this, but few if any other indoor systems do. The ceiling matrix arrangement deviates furthest from it.

(Full planning and installation instructions with box constructional details for the LISCA system are given in the book *Public Address Loudspeaker Systems* by Vivian Capel, published by Bernard Babani, 1990.)

10 Loudspeaker distribution circuits

Maximum power is transferred from one circuit to another when their respective impedances are the same. A high-impedance circuit has a high voltage with low current whereas a low-impedance circuit has low voltage and high current. It follows that if a low-impedance circuit is connected to a high-impedance source, the high voltage will drive an excessive current through the circuit with likely damage to the load or source or both. On the other hand, connecting high-impedance load across a low-impedance source results in low current and low power transfer.

So, for maximum power transfer, the output impedance of the amplifier should equal that of the loudspeaker system, although this is not always the sole consideration with public address systems.

In all cases though, the loudspeaker impedance must be equal or greater than that of the amplifier, never less. A lower impedance would draw high current from the amplifier output transistors, causing overload distortion to start at a lower power level, and excessive heat which could destroy bipolar transistors.

Low-impedance circuits

Domestic hi-fi amplifiers, which are sometimes used for small public address systems, have a low-impedance output of 4–8 Ω. Public address amplifiers have a range of low impedances, usually, 4, 8, and 16 Ω and in additional high-impedance outputs of 70 V and/or 100 V. For small installations it is often more economical to use a low-impedance system.

Total impedance of a low-impedance loudspeaker system must thus always be more than 4 Ω. If there are other loads such as an inductive hearing-aid loop which must also be connected to the amplifier, the loudspeaker impedance must be much higher than this.

When two load circuits are connected to the same source, the power fed to each is equal if the impedances are equal. If they are not, most power will go to the lower impedance. An induction loop usually requires more power than the loudspeakers, so it must be of lower impedance than the loudspeaker system.

A typical example is an induction loop of 7 Ω. To achieve a total above 4 Ω, the loudspeaker system must be 12 Ω. This gives 4.4 Ω and just about the correct power division.

To calculate two impedances in parallel there are two methods:

$$Z = \frac{z_1 \times z_2}{z_1 + z_2} \quad \text{or} \quad \frac{1}{Z} = \frac{1}{z_1} + \frac{1}{z_2} + \cdots$$

For more than two impedances the latter must be used, adding further reciprocals for the extra impedances. Impedances in series are simply added.

The standard loudspeaker impedance is 8 Ω, but 4 Ω units can be obtained and also 16 Ω.

Series–parallel connection

Almost any number of loudspeaker units of any impedance can be brought to near the required value by appropriate series–parallel connection. One useful rule of thumb is that any number of parallel groups, each consisting of that same number of units in series, will give the same impedance as a single unit. For example, five 8 Ω units in series gives an impedance of 40 Ω; and five such groups in parallel gives an impedance of 8 Ω, the same as the single loudspeaker.

Different impedances can be obtained by varying either the number of groups or the number in each group. However, to obtain equal power from all units, the number in each series group must be the same.

For example, a group of six units gives an impedance of 48 Ω. If four groups are placed in parallel, the impedance is 12 Ω (Figure 73). However, if those 24 loudspeakers are divided into six groups of four, each group then has an impedance of 32 Ω, and six in parallel gives 5.3 Ω. The former would be required if an induction loop was also being operated, but the latter would be satisfactory if there were no other load.

If the total impedance is greater than that of the amplifier, maximum power transfer is not achieved. This does not mean that power is wasted, it is just not fully developed in the amplifier. Maximum power transfer is not the sole consideration for small and medium-sized installations. A hall seating 200 probably requires less than 10 W fed to the loudspeakers, whereas most modern amplifiers are rated at 40–100 W, so there is usually plenty to spare. The total load impedance can thus be well above that rated for the amplifier.

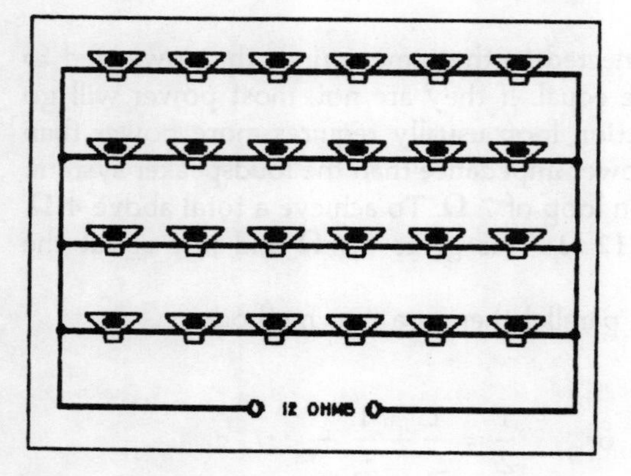

Figure 73 A series-parallel circuit of 4 × 6 8 Ω units produces 12 Ω.

There is an advantage in this, because amplifier distortion usually decreases as the load impedance increases, also the output stage will run cooler, which improves its reliability factor.

The load impedance can be too high, though, in which case the amplifier will provide insufficient power to drive it. Volume controls and mixer then have to be set towards maximum, which could overload the later amplifier stages on strong input signals, so producing distortion.

This is more likely to happen with lower-powered amplifiers, and can be avoided by keeping the load impedance within reasonable limits. As a rough guide, around 5 times could be considered to be the maximum, while a safe minimum is $1\frac{1}{4}$ times, although it can go down to the actual amplifier impedance if necessary. The ideal range is thus $1\frac{1}{4}$–5 times amplifier impedance. The connecting wiring is likely to add about 1 Ω in small low-impedance systems, so this gives a safety margin if the load impedance is close to that of the amplifier.

Most LISCA installations need two arrays and it is convenient, though not essential, if each can be a complete and separate circuit, with its individual cable back to the amplifier. Then either can be run separately for testing. The number of units in each array is the number of tiles across the hall minus four, as those nearest the walls are left free, also a couple near the centre. In most cases the number of units is between 12 and 16.

As there is more than one way to connect an array, the choice depends on the impedance required, which in turn depends on whether one or two arrays are to be used, and whether an induction loop or any other load is to be run from the same amplifier.

If a loop is installed, this can be fed from the 4 Ω tap while the loudspeakers are supplied from the 8 or 16 Ω one. In doing this care must be taken to ensure that not just the load for each tapping, but the total load as seen by the amplifier, does not exceed its rating. If a loop of 8 Ω is connected to the 4 Ω tap, it half-loads the amplifier. If then an additional 32 Ω loudspeaker load is connected to the 16 Ω tap it also half-loads the amplifier, which is thereby fully loaded. Thus 32 Ω would be the minimum limit for the loudspeaker load in this case.

To determine whether the amplifier would be overloaded in any particular case, express the amplifier tap impedance over the load connected to it, as a fraction, and do the same with the second load, then add the two fractions. If the result is greater than unity the amplifier will be overloaded, if less, the loads are within the rating. This can be done also with a third load and tap such as for auxiliary loudspeakers.

Phasing

The importance of phasing individual units in columns has been mentioned in a previous chapter. This is no less true for loudspeakers in complete systems. All loudspeakers working in the same environment must be connected in phase.

If they are not, there will be 'blind spots' in areas where the fields from adjacent out-of-phase loudspeakers overlap.

An exception to this rule is if a single filler loudspeaker is facing in the opposite direction to all the others. It then produces opposite-phase pressure waves and must be connected out-of-phase to correct it.

In order to preserve correct phase, the terminals of each loudspeaker should be clearly marked and the conductors of all twin feeders should be of different colours. With large temporary systems that may be installed in limited time, hence under pressure, it is prudent for all connections to be checked by someone other than the one who made them. Experience has shown that wrong connections are often revealed.

100 V line

All dedicated public address amplifiers have in addition to the usual low-impedance output terminals, one labelled '100 V line'; some have also a 70 V output. Why voltage should be designated instead of ohms, and how in fact a fixed voltage can be specified when the signal is constantly varying, are questions that are often asked. Furthermore, why should anything other than the low-impedance outputs be used anyway?

We will take the last question first. In a large installation, cable runs of several hundred yards or metres may be involved. With twin 13/02 cable having a resistance of 9 Ω per 100 m, it can be seen that if a single low-impedance loudspeaker of 8 Ω is operated on the end of a 100 m run, over half of the power is dissipated in the cable. If another unit is added, the impedance drops to 4 Ω and even greater loss takes place. It may not be practical to place the units in series to increase the impedance as independent volume controls and switches may be required.

The situation where several loudspeakers are needed at the end of a long cable run is a common one, and is quite impractical with low-impedance operation. The answer is to run them at high impedance. Then the resistance of the cable is small compared to the loudspeaker impedance. At an impedance of 2 000 Ω for example, the 9 Ω of a 100 m cable run is insignificant. The circuit impedance is converted down to that of the individual loudspeaker unit by means of a transformer.

Why voltage?

It is of course true that the output voltage from the amplifier is constantly varying with the signal. The 100 V is obtained when the amplifier is delivering its maximum rated sine wave output; likewise with the 70 V output. If the voltage is thus fixed, the actual output impedance depends on the wattage rating of the amplifier. So for a 50 W amplifier, the impedance is: $Z = E^2/W = 10\,000/50 = 200\ \Omega$. A 100 W amplifier has an impedance of $10\,000/100 = 100\ \Omega$.

At first sight this may seem an unnecessarily involved way of specifying an output impedance. It may be thought better to have a standard output impedance such as 200 Ω and stick to it irrespective of the power rating of the amplifier, just as with the low-impedance outputs.

The reason is that a fixed voltage rating makes it more convenient to calculate the power and loading when using a number of loudspeakers of mixed power ratings

The loudspeaker transformers have tapped primaries and secondaries. In the case of the secondaries, these are tapped at the low impedance of 4 Ω and 8 Ω for appropriate connection to the loudspeaker. The primaries are tapped at various wattage ratings and thereby the loudspeaker receives whatever power is designated by the chosen tapping.

As it is the primary that is so tapped, the maximum number of turns are in circuit at the lowest power rating. The highest power is given by the fewest number of turns, hence the lowest tap.

The system is somewhat analogous to the electricity supply mains where there is a fixed voltage, and appliances of different wattage ratings are connected. We do not concern ourselves with the impedance of the appliances, or that of the mains supply, only their wattage ratings and whether our house circuit can supply them.

Calculation of load

To determine whether the load is within the capability of the amplifier to supply, all that is necessary is to add up the various wattage ratings to which the individual loudspeakers are tapped, and ensure that the total is less than the power rating of the amplifier. If it exceeds it, then some units will have to be tapped down to a lower rating to make the figures tally.

Although the output impedance of an amplifier at 100 V differs for different amplifier power ratings, this is of no concern. A loudspeaker with a 100 V line transformer will produce the specified power of the tapping from any amplifier of any power rating.

The system is much more convenient than calculating the total impedance to ensure that it is greater than that of the amplifier, which would be quite a chore with a large installation. Also, widely varying sound levels can be obtained from the same feeder by simply adjusting the transformer taps. This is especially useful in factories, where high volumes may be required in the machine shop, but much lower ones elsewhere. Likewise in a large auditorium, where high power levels are required, while only moderate ones are needed in rest rooms.

70 V operation

We have seen that the output impedance of a 50 W amplifier is 200 Ω. If now we perform the same calculation at 70 V, we get: $Z = E^2/W = 4900/50 = 98\ \Omega$.

So the output impedance of an amplifier at its 70 V tapping is about half as much as it is at the 100 V output.

This means that twice as many loudspeakers of the same impedance can be connected to the 70 V tap as can be to the 100 V, although only half the rated power of the transformer tapping will be available from each. The 70 V output is therefore useful if a large number of loudspeakers are required to operate at low volume and their combined wattage tappings would otherwise exceed the amplifier power rating.

Transformers

100 V transformers are available for a wide range of maximum powers from around a 4 W to 40 W. Each usually has three or four tappings. For example one 4 W transformer has tappings at 2, 1 and $\frac{1}{2}$ W. A 40 W model has 30, 20 and 10 W tappings. For general public-address work a 15 W transformer with taps at 10, 5, $2\frac{1}{2}$, and $1\frac{1}{4}$ W is very useful as it will drive anything from a large column to a mothers' room subdued loudspeaker.

The secondary is sometimes 16 Ω impedance, but is usually 8 Ω, tapped at 4 Ω. Higher impedances can be connected just as with the low-impedance amplifier output. The power is then proportionally less. So a 16 Ω load can be connected to the 8 Ω terminal and the power rating halved. This is useful in the case of columns which, having a series or series/parallel connection of drivers, are usually of higher impedance. The new wattage ratings should then be marked on the transformer to avoid future errors.

Unlike the amplifier low-impedance output, loudspeaker impedances can be taken below the rated impedance of the transformer secondary; in this case the power rating is proportionately increased. So, a column with parallel connected units to achieve tapering, having an impedance of 2 Ω, could be connected to the 4 Ω tap. The power rating would then be double that indicated on the primary wattage tapping. Here again the modified rating should be marked.

If a transformer is required for an unusual load impedance that cannot be accommodated in the above manner, the turns ratio *TR* can be calculated from

$$TR = \frac{E}{\sqrt{(Z \times W)}}$$

It is worthy of note that while care should be taken with all loudspeaker systems to avoid short-circuits, especial care is needed with 100 V line systems. With low impedance, a short at the loudspeaker end puts the resistance of the cable across the amplifier, which is probably a couple of ohms across the 4 Ω output. This may not cause immediate and catastrophic damage if the signal level is not high and the short is soon removed.

With a 100 V line system, such a short places a couple of ohms across an output impedance of 100 Ω in the case of 100 W amplifier. The result will almost certainly be the destruction of the amplifier output transistors unless they have short-circuit protection.

Table 8 Transformer ratios to give specified powers at either 100 V or 70 V working with various loudspeaker impedances

Power at 100 V	1.5 W	3.0 W	5.0 W	7.5 W	10.0 W
1.5 Ω	66.6	47.2	36.5	29.8	25.9
3.0 Ω	47.2	33.3	25.8	21.1	18.3
6.0 Ω	33.3	23.6	18.3	14.8	12.9
7.5 Ω	29.8	21.1	16.85	13.3	11.5
9.0 Ω	27.2	19.25	14.9	12.2	10.5
12.0 Ω	23.5	16.7	12.9	11.0	9.1
15.0 Ω	21.1	14.9	11.6	10.5	8.15
18.0 Ω	19.25	13.6	10.5	8.6	7.45
21.0 Ω	17.85	12.6	9.75	8.0	6.9
24.0 Ω	16.7	11.8	9.1	7.45	6.45
Power at 70 V	0.75 W	1.5 W	2.5 W	3.75 W	5.0 W

Columns and LISCA distribution

In the case of LISCA arrays and conventional columns in medium-sized halls, 100 V operation is unnecessary. The units can be placed in a series–parallel arrangement to obtain a suitable impedance. Cable resistance is only a small fraction of the total load impedance in this case, so although there is some power loss it is quite small. All units are operated at the same power level so the transformer tappings offer no advantage. The cost of providing a transformer for each unit in a LISCA array would also be high. The 100 V system is most effective for large installations, especially where there are a number of individual loudspeakers that may be required to run at different volume levels.

It is a bad practice to connect a large number of low-impedance loudspeakers in series across a 100 V line output, even though the total impedance may match that of the amplifier. An error in connecting, or a fault in the feeder can put a low impedance across the output and damage the amplifier. This can and has happened.

Furthermore there is the possibility of a loudspeaker going open-circuit. Although not common, it does occur, and a large number of units statistically increase the probability. If it does, the whole system goes dead and it is a major job finding the culprit. In a series–parallel arrangement, an open circuit only affects one group, so there is only a partial loss of sound. Furthermore the fault is narrowed down to the affected group and so is easier to trace. This is an example that what is theoretically possible is not always practically advisable.

Auxiliary loudspeakers

In most systems it is necessary to supply low-level loudspeakers in anterooms as well as those in the auditorium. If columns or a LISCA system is in use, this can be run at low impedance while the auxiliaries operate from the 100 V line. These can then be adjusted to the required sound level for each by means of the transformer tappings.

When operating a dual system care must be taken to see that the total load of *both* sections does not exceed the amplifier rating. To do this, first determine how much the amplifier is loaded by the main auditorium low-impedance circuit. Let us say that the total low-impedance load including loudspeaker system and induction loop is 5 Ω, which is connected across the 4 Ω output. This means that the amplifier is running at four-fifths of its load capacity, which leaves one-fifth available for auxiliary circuits.

Translating this into power, if the amplifier is rated at 50 W, then one-fifth or 10 W maximum is available. So, all the transformer tappings of the auxiliary loudspeakers must add up to less than 10 W. This would then fully load the amplifier, but as mentioned earlier it is prudent to under-run it if possible, so a lower figure would be better.

Adding a transformer

It may be necessary to feed loudspeakers in various anterooms from an amplifier that has no 100 V output. In such a case a 100 V line transformer can be installed at the amplifier to convert it to part 100 V line working.

The turns ratio required can be calculated from the same formula as for a loudspeaker transformer:

$$TR = \frac{E}{\sqrt{(Z \times W)}}$$

but in this case Z is the output impedance of the amplifier, which in most cases is 4 Ω, and W is its rated output.

Ratios for popular amplifier powers are from this: 40 W, 7.9; 50 W, 7.0; 75 W, 5.8; and 100 W, 5.0.

It may be possible to obtain transformers designed for these powers from specialist firms, or some transformer winding firms will make one to order although this can be rather expensive.

An alternative is to use a small mains transformer having the same ratios. The low-voltage secondary should be connected to the amplifier output and the 240 V mains primary to the 100 V loudspeaker circuit.

Some mains transformers having the same or near ratios to those required for the popular amplifier powers mentioned above are: 40 W, 30 V; 50 W, 35 V; 75 W, 40 V; and 100 W, 48 V. The voltage in these descriptions is that of the transformer secondary connected to the amplifier. Any other secondaries can be ignored.

The transformer current rating can be quite low. It is unlikely that more than 10 W would be required for the auxiliary loudspeakers, so a 250 mA rating, which gives 10 VA at 40 V, should suffice.

Auxiliary loudspeakers at low impedance

Connecting an 8 Ω loudspeaker directly across the amplifier's 4 Ω output would half-load it and so feed it with more power than the auditorium arrays. If only one to two auxiliary loudspeakers are required these can be supplied by a trick method not generally known, whereby a very low output impedance can be obtained by taking an output from *between* two of the regular impedance tappings on a line transformer.

The secondary of a line transformer starts at a common terminal. The first tap is the 4 Ω, then the 8 Ω and finally the winding ends with the 16 Ω terminal if provided. It may be thought that an output taken between 4 Ω and 8 Ω would be the difference between them, i.e. 4 Ω, and likewise one taken between 8 Ω and 16 Ω would be 8 Ω. This is not so, the windings progress in a non-linear manner, the turns ratio depending on the square root of the impedance.

To find the impedance between two regular tappings the following formula can be used:

$$Z_0 = (\sqrt{Z_1} - \sqrt{Z_2})^2$$

in which Z_0 is the impedance between tappings, Z_1 the higher impedance tapping and Z_2 the lower impedance.

This produces some rather surprising and interesting results. Between the 4 Ω and 8 Ω tappings, the impedance is 0.68 Ω, while between the 8 Ω and the 16 Ω ones, it is 1.37 Ω.

An 8 Ω auxiliary loudspeaker can be run from the 0.68 Ω impedance and will load the amplifier to only 0.085 of its capacity (Figure 74). This is equivalent to a 47 Ω load across the 4 Ω tapping and can be reckoned as such in calculating the total impedance as previously described.

This method can be used when the amplifier is not a regular public address model and has no output tapping. The output impedance can be assumed to be 4 Ω, and a small 100 V line transformer can be used to supply up to two auxiliaries. Ignore the primary, which is tapped in watts, and connect the 0–4 Ω

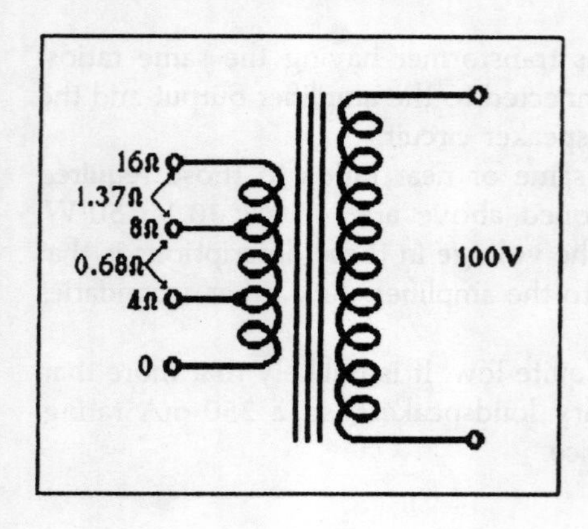

Figure 74 Ultra-low impedances can be obtained between the standard line transformer tappings.

tappings to the amplifier, the auxiliary loudspeaker(s) can then be connected across the 4–8 Ω tappings.

Controls

One or two auxiliary speakers can be switched off or on without affecting the others, but if a large number are installed, some means of maintaining the same load on the line is necessary.

Volume controls can also be fitted consisting of a simple series variable resistor, a potentiometer to load the amplifier when it is turned down and so minimize the effect on other units, or a matching attenuator which provides a constant load at all positions (Figure 75).

The latter uses a two-pole five-way switch giving one full, one off, and three attenuated positions. The resistor values depend on the loudspeaker. For an 8 Ω unit, they are R_1, R_2, R_4 and R_5 2 Ω each; R_3 and R_6, 4 Ω; and R_7, 8 Ω. For a 16 Ω unit these values are doubled.

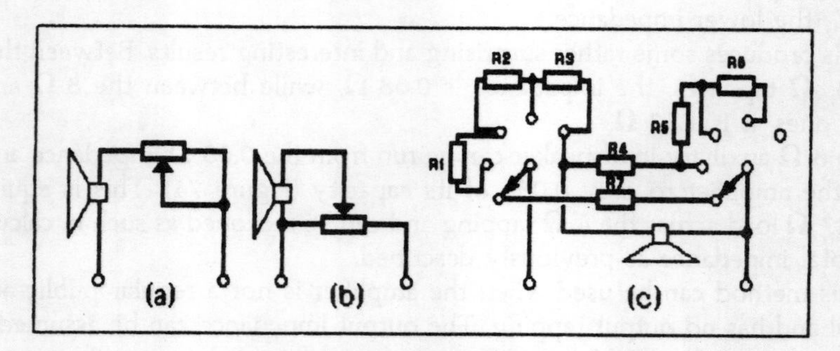

Figure 75 (a) Simple series volume control. (b) Potentiometer maintains load on the amplifier in all positions. (c) Switched constant load in all positions.

Temporary connections

With temporary installations auxiliary loudspeakers can pose a connection problem, especially when a number of 100 V units are to be connected to the same feeder. If the feeder is cut to make a connection at each unit, the end result is a collection of odd lengths of cable, none of which is the right length for the next job.

Various methods of tapping in have been tried with varying degrees of effectiveness, But the best one is the Joyce method. It does not require special connectors but uses items generally available, it makes a secure connection that causes little damage to the cable, has no risk of short-circuiting the feeder or contacting adjacent objects, so producing troublesome earth leaks, and is simple.

A twin screwed-connector is fitted to the loudspeaker lead. A couple of mapping pins are next pushed one through each feeder conductor, separated by the distance between the connector holes. They then are inserted point first into the connector and clamped in place by the connector screws—simple, but effective.

The pins should be of medium length, not too short so that the connector screw only just bites on the point, or too long so as to leave an exposed shank. The screws should not be overtightened as the pins tend to be brittle and can be broken. When piercing the feeder ensure that the pin passes through the centre of the conductor and not just through the insulation (Figure 76). Check polarity after connecting. Mapping pins can be obtained from most large stationers (see Appendix).

When using a reel of feeder that has been used before, and some is left unwound on the drum, make sure that the unused ends are not bare. These may have been connected on a previous job and so left with exposed wires. The possibility of a short circuit is obvious. It is a good practice when using the end of a feeder for a connection, to cut it rather than disconnect it when dismantling. There then will be no possibility of bare wires on the next job.

With large loudspeakers such as columns it may be best for each to have its own dedicated cable which is used on every job. In this case the outer cable end can be bared ready to connect to the loudspeaker, and the inner end brought out through a hole in the drum and terminated with a suitable plug. The feeder can then be paid out from the loudspeaker and the drum with any surplus deposited in the control bay.

A distribution board should be used bearing a number of suitable non-reversible sockets into which the loudspeaker feeders can be plugged. This facilitates testing and location of loudspeaker faults as well as the rapid redistribution of loads if an amplifier fails.

When installing a permanent system, the feeders should be run according to good electrical wiring practice, under the floor or otherwise concealed out of harm's way. With a temporary system care must be taken that feeders do not pose a hazard to the public or are run in a vulnerable position. Doorways can

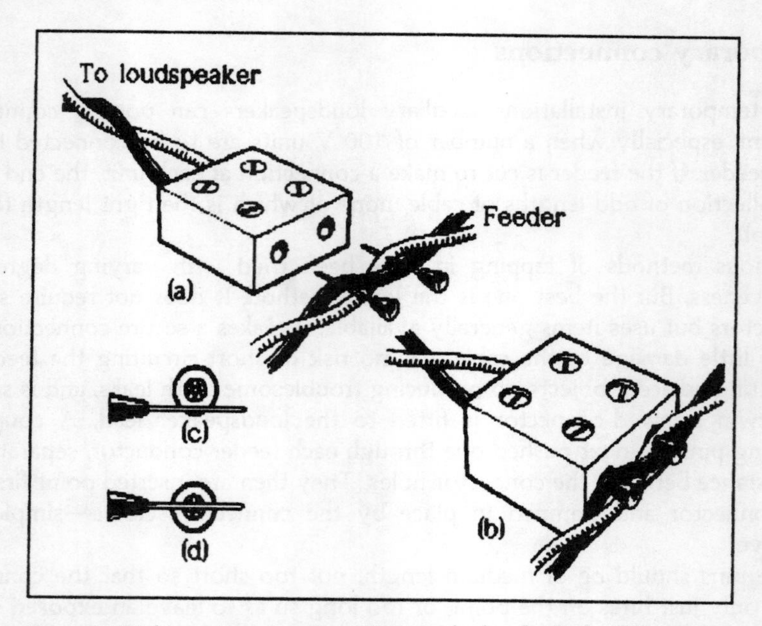

Figure 76 Making a temporary connection to a loudspeaker feeder with mapping pins: (a) push pins through feeder; (b) insert pins in block and tighten; (c) pin through insulation but not wire; (d) pin though the wire. (See Appendix.)

be a problem as a feeder can be pinched and damaged if a door is closed on it, especially on the hinge side. The door should be checked to see where the biggest gap is and the feeder run through at that point. It can be secured with adhesive tape on the frame as this can be removed afterward with no damage to the paintwork. Tape can be used elsewhere as a cheap and temporary fixing agent on paintwork and glass.

Although only temporary, a neat cable run should be aimed for wherever it is visible to the audience. This requires runs that are straight and level, not loose and drooping.

Earth connections

The systems must be connected to a good earth for two reasons. One is screened cables that shield low-level signal circuits from hum are more effective if the screens are earthed. The other is for safety; fatalities have been reported when microphones have become live through some part of the system making contact with the mains. An earthed system is a safe system.

With most small systems the mains via a 13 A three-pin plug is satisfactory. With others, especially large installations, this may not be so. The mains cable often runs for some distance before reaching its earthing point in large buildings, especially outdoor stadiums. It can thereby have small a.c. voltages induced into it as well as other forms of electrical interference. This can be injected into the system.

A separate earth is to be preferred in such cases. Central heating systems can furnish a good earth if the radiators or piping are fixed to a stone or brick wall. The earth wire can be wrapped around a fixing bolt after scraping the metal surface clean. Iron girders, pillars or beams can also serve. With these, a small hole can be drilled at a convenient place and a self-tapping screw fitted as the earthing point.

Recent safety regulations dictate that all units of the system, i.e. mixer, amplifiers and tape-deck, be earthed. If rack-mounted this is automatic, otherwise earth wires should link each one. This may mean that the screens of interconnecting cables may have to be open-circuited at one end to prevent earth loops.

11 Avoiding feedback

Feedback is undoubtedly the biggest single headache that faces the public address operator. Most will have encountered the problem when trying to get adequate gain to amplify a speaker who either mumbles, ignores the microphone or both. The fader is edged up as high as one dares before ringing sets in, but it is still not enough.

Various factors affecting feedback have been mentioned in previous chapters, so we will review these and discuss further measures to counteract it.

Feedback is caused by sound from the loudspeakers being picked up by the microphones, being re-amplified, re-radiated, and picked up again, the process snowballing until a howl rapidly builds up. One of the first considerations, then is to minimize the amount of sound from the loudspeakers that can reach the microphones.

There are two paths that it can take, the direct and the indirect The direct one is a straight line between the two which occurs when the microphone faces the front of the loudspeaker. It is the easiest to avoid.

Although usually placed in front of the speaker's rostrum, the microphone may be moved to any point on the platform for interviews or demonstrations, or if the speaker uses charts or models to illustrate his speech. So all parts of the platform should be out of direct line of sight of the front of the loudspeaker. The worst position for loudspeakers is on the back wall behind the platform. Here they face directly into the microphone, and the path between them is short.

Another poor position is facing downward from the ceiling. This arrangement is bad for several reasons, but a direct feedback path from the first units nearest the platform is inevitable, unless they are well back from the front, in which case the front rows of the audience are not served. With the LISCA system, although the loudspeaker units form a row in the ceiling along the edge of the platform, they are tilted toward the audience. The angle is such that no direct sound is radiated back toward the platform.

One pitfall which could trap the unwary is to put a loudspeaker in front of the microphone, facing the audience. Such a set-up is likely with a small portable system where a single loudspeaker and microphone is used.

It may seem that all the sound is being directed away from the microphone which is behind it, and into the audience. The point being overlooked is that while its rear output is limited by the cabinet back, unless it is airtight, the loudspeaker is radiating from the rear as well as the front. So the microphone is picking up direct sound from the rear of the loudspeaker.

Loudspeakers should be positioned to the side, slightly forward and angled inward to give coverage to the centre of the audience. For columns, a location on the side walls of the hall is usually the most convenient. The inward angle should not be so great though as to bring the microphone within the loudspeakers' field (see Figure 70).

With a wide auditorium, a pair of columns on the platform back wall is permissible if they are placed well to the sides of the microphone. Although the loudspeakers are behind the line of the microphone, the angle from the front of them to the microphone is so wide as to be almost in their region of zero output.

A stage with a facade affords an excellent mounting position as this gives good audience coverage at the centre, yet has no direct line-of-sight path to the microphone (Figure 69, page 137).

Indirect path

The direct path, as we have seen, can be easily controlled by judicious placement of the loudspeakers. A glance is sufficient to see whether an existing system has any direct paths or not, and if so how can they can be eliminated.

The indirect path is another matter. It cannot be completely eliminated in any system that has loudspeakers and microphones operating in the same air mass. The goal must therefore be to reduce the coupling to as low a level as possible.

The indirect path arises from multiple sound reflections around the auditorium. Mostly these are from walls, especially the back wall, but they also are produced from vacant unpadded chair backs, pillars, doors and any other reflective surface. The floor can also produce strong reflections especially from downward-facing ceiling loudspeakers, even if it is carpeted.

As Table 3 shows, the thin carpeting likely to be used for public halls has an absorption coefficient of around 0.25 at mid frequencies, which means it absorbs only 25% of the sound reaching it. Most of the 75% remaining is transmitted through it to be reflected from the floor beneath. On the return transmission a further 25% of the 75% is absorbed leaving some 56% of the original sound to be reflected back into the auditorium. All this reflected sound radiates back to the platform and the microphone.

Just as the careful horizontal angling of column loudspeakers eliminates feedback via the direct path, vertical tilting can reduce indirect reflections. If mounted high and upright as often seen, sound is directed over the heads of the audience to create a strong reflection from the rear wall. The result is reverberant sound and high feedback. If tilted forward as shown in Figure 68, most of the sound is directed where it is wanted, into the audience, and rear-wall reflections are greatly reduced. This can make a significant contribution to feedback control.

The microphone and feedback

Much of the reflected sound can be prevented from entering the microphone if a directional type is used. In order of directivity, the polar responses of microphones are omnidirectional, cardioid, supercardioid and hypercardioid. The omnidirectional instrument is useless for combating feedback because it picks up sound from all angles and so has no significant rejection at all.

The cardioid rejects sound from the rear by some 20–30 dB, but reduces sound arriving at the sides by only 6 dB. Although much better than the omnidirectional, it is not the best, as much reflected sound comes in from the sides. The supercardioid is better, having a side rejection of some 8 dB, but the best is the hypercardioid with a rejection at the sides of 10–12 dB. It has a lower rejection of about 12 dB at the rear due to a small lobe in the polar response, but this is more than compensated by the high side rejection.

It should be noted that many microphones distributors seem hazy as to the polar response of their products. Brochures often describe instruments as simply 'directional'. Others are termed 'cardioid' when tests reveal them to be super or hypercardioid. A polar diagram if published can reveal the true response, but not always, because the concentric level lines are often not calibrated. However, an experienced eye can usually tell the response from the shape of the diagram (Figure 24, page 48).

Another vital microphone characteristic is the flatness of its frequency response. Any peak will initiate feedback. Together with loudspeaker and hall acoustic resonances, the frequency response of a system is shown in Figure 29. With curve (a), the largest peak exceeds the feedback level, so the system will go into violent feedback. Yet the average volume level lies well below the feedback level. To avoid feedback, the gain must be dropped to that of curve (b), and the average gain level is thus shifted even further down.

Peaks in the response thus reduce the amount of gain that can be used before feedback by an amount equal to their amplitude. Furthermore, they are unstable, a sharp peak is far more likely to suddenly trigger feedback when approaching the feedback level than a gentle rise.

It is important, then, to select a microphone that is free from large peaks in its response, as well as being a super or hypercardioid. Often, the only way to determine a microphone's overall feedback rejection is to do a side-by-side comparison with one of known performance.

The criterion in such a test is not the position of the faders at feedback point for each microphone, as this will be dependent on their sensitivities. Rather it is the signal level from the loudspeakers just below feedback, when both the test and the reference microphone are fed the same-level wide-band signal, such as pink noise.

A practical test can be performed using a detuned FM radio which produces white noise, with the two microphones equidistant from it. Each in turn can be taken up to just below feedback level, then the output noted on the amplifier output meter. The one giving the most output is the best for feedback.

The platform back wall

It would seem that as reflected sound arriving at the rear and sides of the microphone is rejected, indirect feedback can be virtually eliminated. Much of it is, but the big snag is the wall behind the platform. Sound from the auditorium can be reflected by it right into the front of the microphone, thereby nullifying all its directional qualities. Being only a few feet away, the reflection is strong and the feedback large.

Fortunately, the solution is not difficult. It is to cover the wall with sound-absorbent material. The chart of absorbents lists a number of possibilities; 4 in (10 cm) fibreglass has the largest absorbent coefficient, but is not very practical. Acoustic tiles or panelling are practical and suitable as they have high absorbent coefficients.

Curtains

One of the best and easiest solutions is heavy curtaining. Curtains are visually attractive as well as having high absorption coefficients. For maximum effectiveness they should hang in deep folds, and a lining also helps. The absorption coefficient is from 0.5 to 0.8 over the mid frequencies, which means that as a sound wave reflected from the rear wall has to pass through the curtain twice, it is attenuated from 0.25 to 0.16 of its former pressure. Further absorption could be obtained if desired by lining the wall over the central area behind the microphone, with acoustic or cork tiles.

The curtains should be from ceiling to floor and extend across the complete wall or at least over a large central area. Random sound arriving from the auditorium comes from all angles. Some can be reflected from the back-wall well to the side of the microphone (Figure 77). So curtains are needed there to muffle it.

Further absorbent treatment behind the curtain need not extend beyond the central area. Reflections from side areas have to twice pass though the curtain sideways, hence through many folds. They are thereby suppressed without need of extra absorbent.

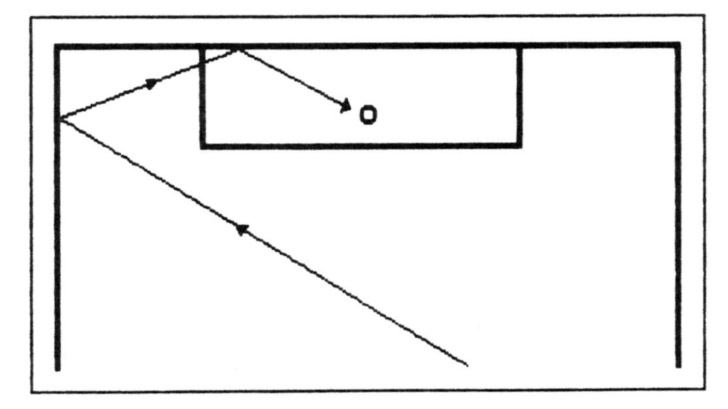

Figure 77 Reflections can enter a microphone from the side wall so absorbent treatment should be extended over a wide area of the back wall.

The beneficial effect of curtaining the platform wall over its whole length cannot be emphasized too strongly. Many installations are known where feedback was hopeless and the purchase of expensive new equipment was planned in the hope of curing it. The advice to curtain the bare platform back-wall was met with some scepticism which turned to astonishment when it was carried out and feedback problems all but disappeared.

Wood panelling

Most of the sound arriving at the platform has been reflected many times from the side and rear walls which are usually plaster-on-brick or plaster-on-cement blocks. A useful reduction of reflected sound from these can often be made by decorating the walls with an alternative less reflective surface. One that has excellent anti-feedback qualities is thin wood panelling suspended on wooden laths. These panels absorb lower frequencies, resulting in a clear acoustic which improves speech intelligibility, and reduces feedback. A wide variety of wood grains and shades are now available, and the visual effect can be very pleasing. There is also scope for making designs or achieving contrasting effects with different woods.

Panelling could be used for the platform back wall, and would certainly be better than bare plaster, but as the wall plays such a significant part in feedback it really needs the much greater absorption of a heavy curtain.

Concrete pillars, columns and ceiling beams may cause strong reflections back to the platform at higher frequencies. Any that are near the front of the hall and are more than a few inches wide could be covered on the side facing the platform with a convenient absorbent.

Loudspeaker phasing

It is sometimes found that feedback is less with the loudspeaker feeders connected one way round than the other. This is not always so, as the random nature of feedback reflections usually make loudspeaker polarity of little consequence. It is always worth trying, though; in some cases the difference is appreciable.

Structural feedback

Indirect feedback can occur from a loudspeaker that is standing on the floor, back through the flooring to the microphone stand. This can happen with the heavy stage-standing loudspeakers used by groups. It can be readily diagnosed by holding the microphone stand clear of the floor. Thick pads under the loudspeakers should cure it, and/or a pad under the microphone stand.

Frequency shifters

Having considered how feedback can be reduced by the choice and correct installation of microphones and loudspeakers, and suitable acoustic treatment, we now take a look at some electronic devices for controlling feedback. The first of these is the frequency shifter.

Frequency shifting is an effective way of reducing feedback, though not without its disadvantages. It works by raising all frequencies that pass through it by about 5 Hz. For speech this has no audible effect as the percentage change in pitch is too small to be noticed. However, the feedback signal does not pass through the system once, but thousands of times, rapidly. Each time round its frequency is raised by 5 Hz, so it does not reinforce itself and is quickly shifted away from the resonant peaks, being soon pushed to above the audible range.

The sound when the gain is taken up to the feedback level is that of a succession of bleeps of ascending pitch, like some car alarms, rather than the continuous howl normally obtained.

One type of frequency shifter works as follows:

The input signal is split into two paths and by the choice of the values of coupling components the two signals are in quadrature at all frequencies (90° out of phase). A quadrature sine-wave oscillator running at a frequency of 5 Hz has two outputs that are also 90° out of phase.

Each audio signal is multiplied in a quadrature mutiplier, by a different 5 Hz output. This produces sum and difference signals in each path consisting of the audio plus 5 Hz, and audio minus 5 Hz.

With the minus signal, we have a −90° phase difference between the two audio signals to start with, and a +90° difference between the two oscillator outputs. When subtracted these produce a total shift of −180°, so that the minus signals in the two paths are out of phase.

With the sum signals there is +90° signal difference between the audio signals and a −90° difference between the oscillator outputs. These, when added, give a zero difference so that the two plus signals are in phase.

The two paths then are combined so that the out-of-phase difference signals cancel each other, whereas the in-phase sum signals reinforce. The output thus consists only of the sum signal of the original audio plus 5 Hz.

Careful adjustment of preset controls is required to ensure the circuits are balanced and the two signals are of equal amplitude, also the oscillator must be accurately adjusted.

The device enables up to 6 dB extra gain to be obtained before feedback, which is very useful. A major feature is that it requires no setting up in the system it is to operate with, and is not affected by changing acoustic conditions.

This is especially advantageous with quick set-ups with which there is little time to make lengthy adjustments before going 'on the air'. All that is required is to connect it between the mixer and the amplifier and switch on. The internal adjustments should stay good for some time once they are made, although they may require occasional re-setting.

The 5 Hz shift has no audible effect on speech, but it does have on music. If the shift resulted in a small uniform change of pitch over the musical scale there would be no problem, but being a fixed frequency, the proportion changes. For example, middle C has a frequency of 261 Hz, and C#, which is a semitone higher, has a frequency of 276.5 Hz, a difference of 15.5 Hz. A 5 Hz shift is thus a third of a semitone sharp here.

At two octaves below, C_2, the frequency is 65.4 Hz and C# is 69 Hz, a difference of 3.6 Hz. Here, a 5 Hz shift is equivalent to well over a semitone. At C_3, there is only 4 Hz difference between C and D, so 5 Hz is more than a whole note sharp.

If music is played through a frequency-shifter, the result can be quite inharmonious and sound as though the bass players are playing all wrong notes. So the frequency shifter is not suitable for groups, or other live music performers other than for special applications.

The same effect is obtained with recorded music, but anti-feedback measures are not required for records so a frequency-shifter would not be needed anyway. If a public address system has provision for recorded music as well as speech, and a frequency-shifter is used, the tape or disc input should be designed to bypass the shifter.

A phase shift of exactly 90° at all frequencies may not be obtained, so the minus signals may not fully cancel. The result could be a beat between the two, and amplitude modulation. Although this is likely to be only a small amount, it is a further degradation of the signal in addition to the disproportionate pitch changes over the frequency spectrum.

The effect of peaks

Referring back again to Figure 29, we have seen that the amplifier gain is restricted by the peaks in the system. The overall level cannot be raised any higher when the tips of the peaks reach feedback level. So to obtain stability and stop ringing or *incipient feedback* as it is termed, the gain must be reduced so that no peak encroaches on to the feedback line.

So, as we have seen, microphones should be chosen that have as flat a frequency response as possible. However, all microphones, other than studio capacitor models, have some variation in response, hence small peaks. Loudspeaker cones also have fundamental resonances that cause peaks in their response, and both these and those of the microphone, have harmonics at twice, three times, four times, etc., the frequency of the fundamental. Added to these are troughs and peaks which are the result of comb filter effects due to reflections or multiple loudspeaker sources in the auditorium.

The result is a frequency response that is far from flat, but consists of a series of peaks. Feedback occurs at the frequency of the largest peak, which is the first to cross the feedback line when the gain is raised. If these peaks or even the

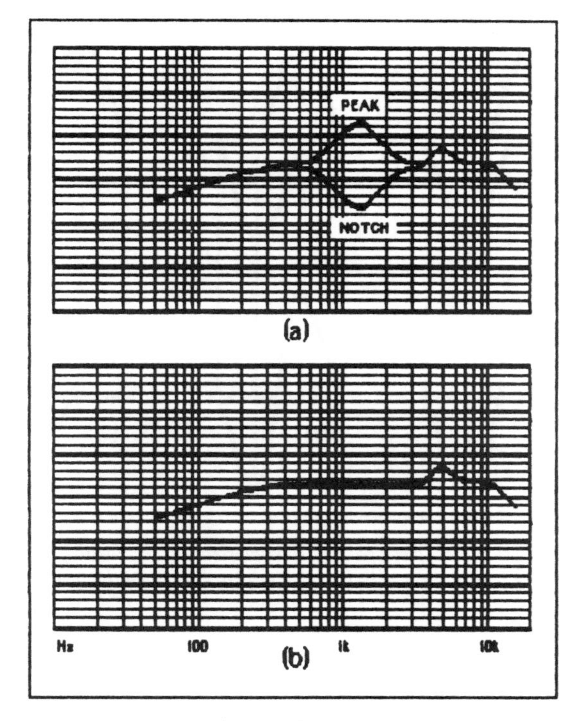

Figure 78 (a) A notch is produced that is the inverse of a feedback peak. (b) The peak is eliminated leaving smaller peak and reduced feedback.

largest could be eliminated, then the gain could be brought up to a much higher level before any part of the response curve encroached over the feedback line.

Notches

One way of eliminating a peak is to superimpose on it its exact inverse, which is a notch of the same frequency and amplitude. This is shown in Figure 78 (a) and the resultant obtained by adding the two is shown at (b), which is a level response. The notch must be of exactly the same frequency or very close to it. If it is displaced by more than a half the width of the peak, it has no effect on the amplitude of the peak at all.

The effects of displaced notches are shown in Figure 79. In (a) the notch is displaced by threequarters of the width of the peak. The resultant shows that the peak has a sharper fall-off but is undiminished in amplitude. A similar effect is shown (b), where the notch is displaced by half the peak width; the fall-off is sharper, but the peak amplitude remains the same. In (c) the notch is displaced by a quarter, and although the peak is reduced by some 40% it is still not eliminated.

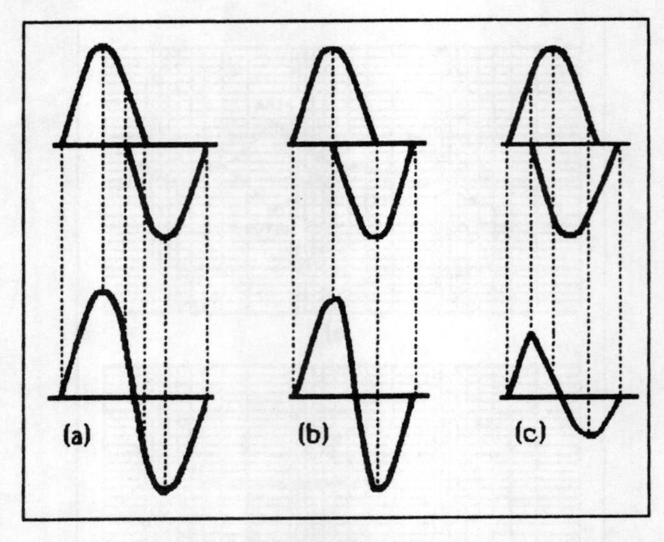

Figure 79 (a) Notch displaced by threequarters of the width; (b) notch displaced by half of the width; (c) notch displaced by a quarter of the width.

Another factor is the peak's width. When a physical object resonates it is not just one frequency that is emphasized, adjacent ones also are affected although to a lesser degree. The fall-off on either side of resonance is usually at a rate of 6 db per octave. If two resonances coincide, the fall-off is at 12 db per octave, but this is uncommon.

If a notch of exactly the same frequency but with a sharper fall-off hence of a narrower width than the peak is applied, two smaller peaks are created, one on each side of the original (Figure 80 (a). They are not affected by increasing the amplitude of the notch, which merely deepens the wall between them (Figure 80 (b). If the notch is not exactly the same frequency, the peaks will be of unequal size.

Another possibility is where there are two notches, one either side of the peak. If they are spaced so as to just overlap the peak their only effect is to narrow it. But when they are close, they merge to form a double-dip notch, and this can reduce the peak amplitude. If the notch depth is increased it can eliminate the peak entirely; however, a large area of the surrounding spectrum is thereby engulfed, so affecting the tone. From all this it is evident that the only satisfactory way of eliminating a peak is by a notch of exactly the same frequency, amplitude and width.

Graphic equalizers

Graphic equalizers are sometimes used to try to eliminate peaks. These have a number of controls, usually sliders, each controlling a narrow band of frequencies.

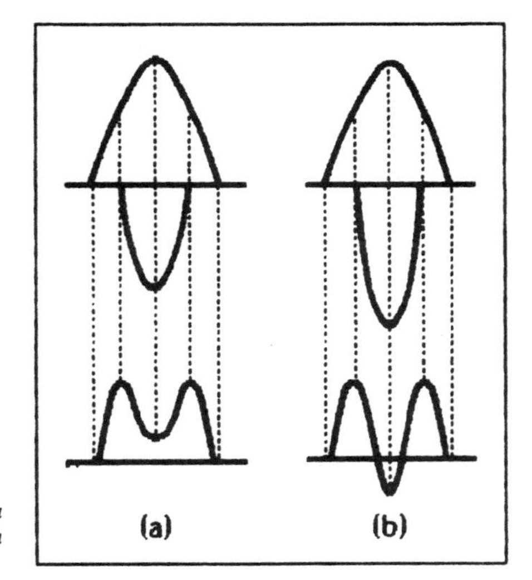

Figure 80 (a) Two smaller peaks are produced by a notch of insufficient width. (b) Increasing notch depth leaves peaks unaffected.

When advanced above the centre level position, the control gives a boost to that particular band, and when it is lowered, the response is reduced. It thus produces either a peak or a notch in the overall frequency response. It is termed *graphic* because the settings can be readily seen from the positions of the sliders.

The audio spectrum is divided into ten octaves. Each octave is twice the frequency of the one below it, so starting from the bottom, and with figures rounded to the nearest whole number the ten octaves are: 16–31 Hz; 31–62 Hz; 62–125 Hz; 125–250 Hz; 250–500 Hz; 500–1000 Hz; 1 kHz–2 kHz; 2–4 kHz; 4–8 kHz; 8–16 kHz.

The simplest equalizer has five bands, which means that there is only one band to two octaves. Such an instrument can do little more than enable a subjective adjustment to be made to a music system, such as one installed in a car. The chance of one of the bands coinciding exactly with a resonance peak is remote, so it is virtually useless for public address.

A more common domestic equalizer is the ten-band model. This provides one control for each octave. For stereo this has to be doubled to twenty bands, but only mono is required for public address. Again the resolution is insuffi ient. The chance that a feedback peak would exactly coincide with one of the equalizer control frequencies is only twice as great as for the five-band model. The bands are too widely spaced to produce an effective double-dip elimination of a peak by adjacent bands, and any attempt to do so would take out a large chunk of the spectrum.

The equalizer used by professionals is the third-octave type, having thirty controls per channel. There is much better chance of a band coinciding with a peak, but even with this chances are less than even of getting it exact. Should it happen, the notch may then be narrower than the peak, depending on the model. However, there is a fairly good chance of taming the peak with two adjacent bands.

Setting up an equalizer

To start, all controls should be at the level position. The volume is turned up until feedback occurs. Then the controls which affect the peak must be discovered. This is a matter of trial and error, which with thirty to choose from can be rather tedious and time consuming. However, the choice can be narrowed with judgement, which tends to improve with experience.

If the feedback note is a hoot, try the controls from 250–500 Hz. If it is a singing tone, try around 1 kHz, but if it is a whistle, go for 2 kHz and above. It is unlikely to be in the top octave 8–16 kHz, or the bottom two, 16–62 Hz.

Lower the controls in the suspected region one at a time, restoring each afterward until the feedback level changes. If more than one affects it, find the one that affects it most, as this will be the nearest channel to the peak. Turn the slider down until feedback stops and then turn up the system gain control. If feedback starts again at the same frequency take the slider further down or try an adjacent one.

Adjust the settings and that of the gain control until the feedback changes pitch. This means that the primary peak has been reduced and the next largest has taken over. Repeat the operation for this one.

It may then be found that the first peak returns. This is because it wasn't completely flattened but just reduced to a lower level than the second. When that one is lowered, the remnant of the first peak takes over again as the largest. So further juggling and adjusting is required of both, during which a third tone may put in an appearance.

The setting up obviously requires patience, and if done in an empty hall will probably need readjustment when the acoustics are changed by the arrival of the audience or when the temperature changes. Added to these drawbacks, thirty-band equalizers are not inexpensive! But there is a better way.

Variable-notch filters

A variable-notch filter is a device that produces a notch but no boost, which for this purpose is not required. The major advantage over an equalizer is that the notch is continuously variable and so can be turned to exactly the same frequency as the peak. Two notches are generally sufficient, each with a tuning control, but in theory any number could be cascaded. The device is smaller, less formidable, much easier to set up, is cheaper, and because it can mirror the peak exactly, is more effective than an equalizer.

Apart from reducing feedback the notch filter has another useful feature. Reproduced reverberation through the system robs speech of its clarity. This tends to be greater at frequencies around the resonance peaks, especially when the gain is up near the feedback level.

As the notch levels out the peak, it also reduces the reverberation coming through the system. In fact, if the notch is tuned through the peak when a programme is in progress, the improvement clarity becomes quite noticeable.

The filter thus restores natural tonal balance by removing spurious peaks, and also reduces confusing reverberation which centres around those peaks. So as well as giving more gain before feedback, the notch filter cleans up the sound of speech and generally improves it. These advantages are not obtained with a frequency-shifter. Its action could be said to skate over the peaks, thus preventing them causing feedback, but they remain to colour normal non-feedback signals, and reverberation is likewise not reduced.

Another advantage of the notch filter is that feedback is initiated less readily when the feedback level is approached by a flat response than by a sharp peak. It behaves like a well-damped oscillatory electronic circuit. If there is no peak, when feedback level is exceeded the circuit doesn't know at what frequency to oscillate. It seeks any small remaining peaks and starts to oscillate slowly and sluggishly.

In contrast a sharp peak will start things oscillating smartly, and the operator must be ever vigilant. So it is much less nerve-racking to operate with a flat response than with one with peaks!

The device is very easy to set up. System gain is advanced until feedback occurs, then the filter is turned to eliminate the peak. The peak depth control (if fitted) is adjusted so that the peak is just cancelled so that the response is then flat at that point. If a twin filter is used the second largest peak can be dealt with in the same way by the second filter. With the two largest peaks exactly cancelled, the remaining ones, being much smaller, have little effect.

The amount of extra gain available by using notch filters depends on the peaks in the system. Greatest benefit is obtained when peaks are large, as their elimination brings a bigger increase in usable gain. Conversely, when the peaks are small there is less increase. However, all systems have peaks, so some improvement is certain, and the extra clarity and slow feedback characteristic makes the unit a very useful tool in the fight against feedback.

(A full description with circuit layout and constructional details of a twin-notch filter with depth control can be found in *Acoustic Feedback, How to Avoid It* by Vivian Capel, published by Bernard Babani, 1991)

12 Planning reliability

An important public meeting has been arranged to hear a visiting celebrity speaker. Large sums have been spent in advertising and hiring a hall seating several thousand, which is packed to capacity. Many have travelled long distances to be present. Suddenly, just after the start of the proceedings the public address system goes dead.

Frantic efforts to restore the sound fail, and only the first few rows hears anything of the rest of the meeting. This could well be described as the public address operator's nightmare, but it can happen, especially with a temporary installation in which there are so many possible reliability hazards.

The only way to reduce the possibility of trouble and to quickly deal with it if it should happen. is to build reliability into the system. Identify potential trouble spots and eliminate them as far as possible.

Standby equipment

Power amplifiers are the most vulnerable part of the equipment as they are at the mercy of any fault on the loudspeaker feeders, they handle large powers, and they get hot. As all electronic maintenance engineers know, wherever there is heat there is a high incidence of failure.

Amplifier capacity should always be greater than that actually required. This can be either by the provision of a standby amplifier not normally in use, or by underloading the amplifiers that are used. In either case some means of quickly re-distributing the loads must be arranged. A practical method is a loudspeaker feeder distribution board consisting of several groups of non-reversible sockets. Each group is connected to one amplifier, and has a number, perhaps four, sockets so that several feeders can be plugged in to the same amplifier.

With large installations a label can be attached to each feeder, or it can be colour coded and an information chart kept nearby. This should record the area it serves and the total power tappings of all loudspeakers connected to it. It can then be seen at a glance how fully the amplifier is loaded and what spare capacity it has (see Appendix).

Another useful item that can be recorded is the impedance of the feeder. This should be measured with an impedance meter after the initial installation and testing is complete. If no impedance meter is available, the d.c. resistance can be taken and recorded instead. Although this is not as informative as the impedance, it can still be useful. Any line fault subsequently occurring can then be quickly identified by comparing the reading with the original.

The output of the amplifiers needs to be monitored. Output meters give an immediate indication that all units are functioning, but they do not indicate the presence of faults such as noise or distortion. Some means of audible monitoring is therefore desirable. A cabinet loudspeaker with internal line transformer can be fitted with a short lead and plug, to plug into any vacant socket on the distribution board. A periodic listening test to all the amplifiers in turn can thereby be made, but if all sockets for a particular amplifier are occupied, it cannot be tested.

A better arrangement is to mount a loudspeaker on the amplifier rack with a multi-position switch which is connected to all the amplifiers. The test is then made by simply switching around. A refinement is to have a volume control for each position, to be set so that all give the same volume on the monitor irrespective of the amplifier control settings. Should any amplifier suffer a drop in level, this will then become apparent when switching from the previous one.

The mixer, although complex, is less vulnerable and rarely gives trouble, but a simple standby with two or three microphone inputs and a line input is worth having available. It may not have all the facilities of the main one but it could keep things going if that one failed. It could also be useful as a diagnostic substitute in the event of an obscure fault.

A mixer should be chosen that has a headphone monitor facility, and of course a pair of headphones should be on hand to use with it. The operator should wear these at all times, so that the quality and level can be constantly checked. Should any fault occur it will be immediately evident whether the trouble is in the mixer or elsewhere.

Microphones

Microphones are very reliable and stand up to hard usage. Ribbon transducers are rather fragile and need care, but even these do not usually fail unless subject to physical damage. Electrets rely on internal batteries so these should replaced at regular intervals even though the current drain is small and life can be several years.

Noisy contacts are likely to be the main problem with electrets. A drop of oil on the switch followed by rapid operation should be applied once in a while. This cleans the contacts and prevents future tarnishing. Battery contacts should be lightly greased.

A spare microphone ready connected, with its lead and plugged in to a spare channel on the mixer, should be available in a handy position on the platform. Should a microphone or its lead give trouble little time will then be lost in replacing it. The spare can also serve as a quick test as we shall see.

Loudspeakers

Loudspeakers can go open-circuit although this does not often happen. The trouble usually lies with the soldered joint on the cone connecting the end of the coil winding to the flexible wire that goes to the terminal strip. If this happens all loudspeakers connected in series with it are silenced. So, while series arrangements are necessary to obtain the required impedance, it is not wise to have too long a series chain, certainly not to have all the units in series.

A series–parallel configuration will serve most of the audience if one series group fails. If possible, every section of the audience should by served by units from more than one group. Then all will be served in the event of a failure, although at reduced level. Thus in a LISCA system alternate loudspeakers could be of a different series group.

Cables and feeders

Undoubtedly, more trouble arises from the cables and feeders than anything else, especially in a temporary set-up. If due consideration is given to these, their choice, care and maintenance, many potential causes of breakdown can be averted.

Microphone cables can be a prime cause of trouble. They are subjected to much flexing and bending, metal fatigue eventually takes its toll and the conductor breaks within the insulation. In nearly every case the break is within a few inches of the microphone itself as this is where most of the flexing takes place.

To reduce the flexing, cable connectors are normally fitted with a strain reliever in the form of a rubber or plastic tube with walls of decreasing thickness. This distributes the bending over the length of the tube instead of at one point immediately beyond the exit. In many cases though, all this does is to shift the breaking point to just beyond the end of the tube.

The best insurance against cable breakage is to use the right cable. Single screened cable is the most vulnerable and should not be used for microphones even with an unbalanced input system. Twin twisted screened cable with padding is the most suitable type, as it resists sharp bending. Further immunity from breakage is achieved by using 16/0.2 conductors rather than the smaller 7/0.2 size.

An even better cable is twin twisted padded with two contra-lapped screens for maximum screening and 55/0.1 conductors for maximum flexibility. However, this is expensive and is not really necessary for public address use, unless there is an extremely noisy electromagnetic environment. It is more suitable for studio work.

For balanced systems twin cable with be necessary anyway, but for unbalanced ones there is a bonus in that the quasi-balanced mode described in Chapter 4 can be employed.

When microphone cables are reeled up or unreeled by hand, twisting is inevitable unless the reel is rotated as the cable is laid on or pulled off. The result is unsightly loops and possible damage. The alternative is to store the cable on rotatable plastic drums such as commonly used for mains cable extensions. A short length of cable can be brought out from the centre of the drum and the mixer plug fitted. Thus only the required amount of cable need be pulled off the drum which can then be deposited near the mixer. This affords a neat and tidy set-up which is easy to lay out and take up again, and which is the least vulnerable to damage.

The other big hazard is the loudspeaker feeders in temporary systems. When running through doorways they should be taped to the door frame at the point where there is the biggest gap between door and frame. This is usually the top corner on the opening side; there are rarely adequate gaps on the hinge side.

As far as possible, all feeders should be well out of public reach so a high-level route is preferable to a low-level one, e.g. along the floor. Corners should not be bridged but the feeder should be taken right into them. More than one feeder has been brought down from across a corner by a ladder or pair of steps being placed against it.

Surplus wire should not be left out in the auditorium. The feeder run should start with the furthest loudspeaker and be brought back to the amplifiers, any surplus being left on the reel in the same way as for microphone cables. Care should be taken with tappings made from the feeder to ensure that the conductor is penetrated by the pin and that the polarity is correct (see Figure 76). These connections should be checked by someone other than the person who made them.

Sometimes a length of wire may have to be joined to make a longer feeder in a temporary installation. As feeders often have to withstand strain and stress a good physical joint must be made as well as an electrical one. As good a method as any is to tie the wires in a reef knot so that the two free ends can be connected, insulated, and taped back to the feeder (Figure 81). There is thus no strain on the connections.

If such a join made on a previous job is encountered, it is a good practice to cut it out and remake it. Unless soldered, twisted or tied wire joins can become loose or tarnished in storage and so may give rise to an intermittent connection. Intermittent faults are the most difficult to trace as they never seem to go wrong when you are testing for them. Extra care to prevent them is clearly well worth while.

The lead from the mixer to the amplifiers is not particularly vulnerable as it remains in the same position within the operating area and is thereby always

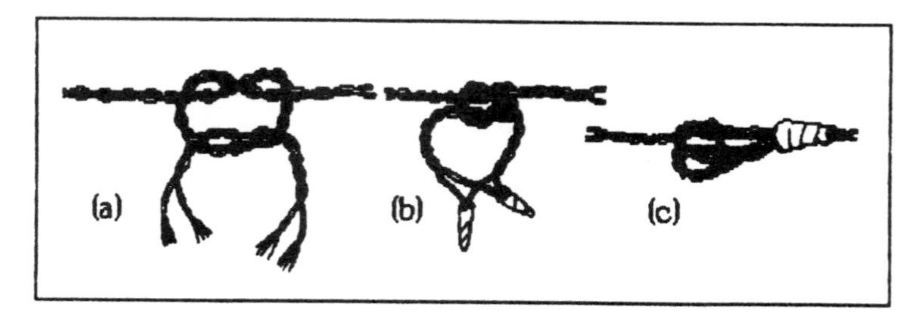

Figure 81 Jointing a loudspeaker feeder: (a) knot is tied; (b) connections are made and insulated; (c) ends taped back.

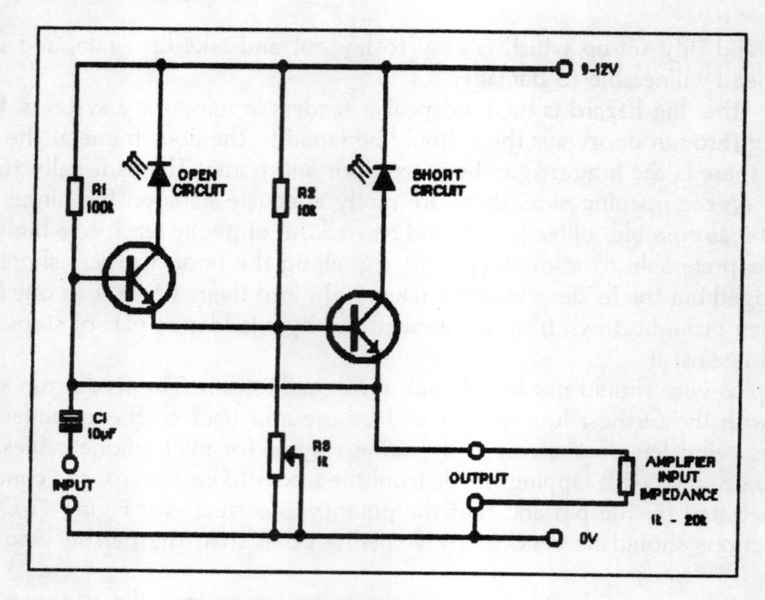

Figure 82 Signal-cable fault warning circuit.

under the care and control of the operator. However, it is the one cable that carries everything, so if it goes everything goes. It could be inadvertently pulled, thereby breaking a connection in the plug, or previous damage could go unnoticed but choose to reveal itself at some crucial point in the programme. It has happened! A spare lead should therefore always be handy and ready to use.

The circuit of a cable-fault indicator is shown in Figure 82, which can be built into a mixer or mounted separately. It is especially useful where a mixer may be operated in a remote position from the amplifiers, so requiring a long vulnerable connecting lead.

Indication is by two LEDs: one lights if the lead goes open-circuit, the other if it goes short-circuit. Values of the resistors may need modifying to suit the H_{FE} of the transistors or for a 12 V supply. The effects are as follows: R_1–no effect on brightness of either light emitting diode but the open-circuit LED lights when the output is loaded if the value is too low; R_2 decreasing the value makes the short-circuit LED brighter, no effect on the open-circuit LED R_3—increasing value brightens the short-circuit LED but darkens the open circuit LED; adjust for equal brightness.

The load is a nominal 10 kΩ but must lie within 1–20 kΩ. If the input impedance of the amplifiers is too high, a resistor should be shunted across the combined inputs. C_1 blocks d.c. from the output circuit of the mixer, while the input capacitors of the amplifiers should block d.c. there.

Power supplies

Just as vital as the mixer lead is the power supply. Most mains supply outlets are now part of 13 A ring circuits, with earth trips in place of fuses. Should some part of the public address mains distribution system develop only a small leakage to earth, the circuit will trip. If this should happen much time could be lost finding the right trip from the maze of boxes and meters that usually adorn the electrical bays of most public buildings. The solution is to find it in advance; check and note which one controls the socket that supplies the public address system.

A supply of spare plug fuses should always be available, but the plugs themselves are not above suspicion. Connections and tightness of the fuse in its holder should be checked with each plug used. Any sign of warmth from any of the pins after use should be immediately investigated, as this is a symptom of arcing due to a loose contact.

Total failure of the mains supply during a large event is not a common occurrence in a city area, although it has happened. It is more likely in rural districts, especially in bad weather. In such conditions some thought as to the possibility would not be amiss. A battery/mains amplifier available as a standby could be a prudent insurance. It would need to have at least a couple of microphone inputs as the mixer would also be out of action. For larger events, two or more such standbys would be needed with sockets to couple one to another as a slave.

Some plan as to the battery supply would also have to be worked out. Keeping and maintaining batteries for the purpose may not be economical as the need to use them may rarely if ever arise. One solution is to get a couple of volunteers to park their cars nearby and be ready to remove their batteries if required. Suitable battery leads for connecting to the amplifiers should not be overlooked. There could be no objection to the short delay incurred as a total mains failure is hardly the fault of the public address system. Some delay would be inevitable anyway to change the amplifiers and all connections thereto.

Another option is the transverter, which enables a 240 V a.c. supply to be obtained from a 12 or 24 V input. Models are available that give powers from 50 to 500 W. The big advantage here is that no alterations are needed to the system or its connections, and the existing amplifiers and mixer as well as tape or other inputs can be used as before.

The higher-power output transverters take a heavy drain from the batteries at full load, so prolonged running may not be possible. There are also conversion losses of about 14%; a 500 W supply at 24 V thus means a drain of some 24 A. However, power taken is related to the power actually used, so if a high-power transverter supplies only low power, the battery current is in proportion. As class B amplifiers only take their maximum current on sound peaks, the average current is less than may be expected from the amplifier power rating. Standard car batteries should thus be capable of maintaining operation for a reasonable period, though not indefinitely.

There are two main types of transverter, one generates a square-wave output which is most suitable for machinery. The other generates a sine wave, which should be used for amplifiers.

Breakdown drill

What happens if there is a breakdown? The procedure should be well thought out in advance and a drill established. The first rule is not to panic, (at least not at first). Not easy, this, if there are a thousand people waiting for you to restore the sound and you haven't a clue what the trouble is!

The first objective is to establish the nature and location of the fault, the second is to decide what to do about it. The immediate action is trace the cause by a process of systematic elimination. Most common faults are: loss of signal; a drop in signal level; distortion; and noise. Any of these may be continuous or intermittent.

The first instant check is whether the fault is evident on the mixer headphones, if it is, fading up the standby microphone and the main one down will reveal whether the fault is due to the microphone, its lead, the mixer input channel or later stages of the mixer.

If all is well at the headphones, a check of the amplifier monitor will indicate if the trouble is on all amplifiers or only one. If only one, this can be taken out of service and its load redistributed among the others. If all are affected the mixer lead would seem to be the culprit. If the cable monitor previously described is in use, this would be immediately apparent from the commencement of the fault.

Instability can cause particularly baffling symptoms which seem to defy systematic diagnosis, but more of this in Chapter 18.

13 Induction loops

A growing number of public halls are having induction loops for hearing aids installed in them. Once they were a novelty, but are now becoming as common and accepted as the public address system itself. Undoubtedly they confer a major advantage to the hard-of-hearing.

Sound from a public address system sounds indistinct, boxy and distant when picked up with a hearing aid. This is partly due to monaural hearing—being heard with only one ear. Normal hearing with two ears, subjectively partly compensates for hall reverberation, which is mainly responsible for these effects. Also, nearby sounds, such as coughs, whispers, programme rustling, and babies crying, sound much louder in proportion when heard through a hearing aid.

When receiving the programme from a loop, the result is as if the earpiece is connected directly to the speaker's microphone. The sound is loud and clear, hall acoustics are minimized,and extraneous noises are hardly heard at all. There is no need for special seating areas, the programme can be picked up anywhere in the hall, and often in adjacent anterooms as well.

It is a fairly straightforward installation to make, as it consists of simply running a loop of cable around the perimeter of the hall and connecting it to the amplifier, but there are several factors to be considered. These are: the gauge of the cable, the number of turns in the loop, its position, matching its resistance along with other loads to that of the amplifier, and the power available from the amplifier.

First of all, it must be noted that the signal is only picked up by hearing aids that have a two-position switch. One position is marked M, and is the normal one by which the internal microphone is connected. The other is marked T, which introduces a small coil into the circuit.

Originally, this was intended to pick up magnetic fields from telephone handsets, thereby giving a clearer result when using the telephone. All NHS hearing aids made since 1974 have this switch, but many private models do not. These are the ones that fit totally inside the ear and so do not have room for a coil and switch.

Field strength

The ideal strength is that which presents a signal to the hearing-aid which is comparable to the output of the internal microphones. Too weak a signal is not desirable as this means the user has to turn up the gain of the hearing aid, thereby making the noise of the internal amplifier noticeable. The British Standard BS 6083 Part 4: 1981 specifies the optimum strength as 100 mA in a single-turn loop of 40 in (1 m) diameter.

Basic to all calculations is the principle that it is current and the number of turns that influence the resulting field in any given size of loop, not voltage. Negligible power is extracted from the loop by the hearing aids, so the voltage required is only that needed to drive the current through the loop. If the resistance can be made very low, the necessary current can be achieved with only a small voltage. However, as the magnetic field is proportional to the product of the current and the number of turns, it can be an advantage to increase the number of turns even though this also increases the resistance. Resistance can be reduced if necessary by using a heavier gauge cable.

The specified current of 100 mA/metre diameter is for an average signal level, but peaks will exceed this by several times. The British Standard recommends allowing for peaks of 12 dB above average which increases the current requirement by four times. If dynamic range compression is used in the amplifier, this can be reduced.

When the signal is mainly speech, a lesser allowance for peaks may be adequate. In practice, an allowance of 6 dB has been found to be sufficient, but to provide a reasonable margin, the calculations that follow assume peaks of 10 dB or three times the average.

With a required average current of 100 mA per meter and a loop diameter d, the current is therefore $d/10$ A, and the peak current $3d/10$ A ($d/11$ and $3d/11$ if d is in yards).

This is the case for a circular loop. A square loop needs slightly more current to provide the same field, about 112 mA for a square having sides of 1 metre, so the formula for peak current becomes $3d/9$ A ($3d/10$ for d in yards).

However few halls are square, most are rectangular. To find the exact figure for the peak current is rather complicated, but an approximate figure for a hall that has a length that is no more than $1\frac{1}{2}$ times the width can be obtained from its area A. The formula is

$$I = \frac{\sqrt{A}}{3} \quad \text{or} \quad \frac{\sqrt{A}}{3.3}$$

for A in metres or yards respectively.

In the case of long, narrow areas things are rather different. With a square loop, each side contributes equally to the field, but if a small square section somewhere near the middle of a long narrow loop is considered, the short sides are too far away to have much effect, and only the central parts of the long sides are generating field within the section. Hence the field is approximately half what it would be for a square loop of the same width. At the ends of the rectangle three sides are contributing, so the field is about three-quarters of a similar square. Field strength thus varies over the length, being greatest at the ends.

Turns

So far we have considered only single-turn loops. These in practice are not very efficient, requiring large currents to produce the required magnetic field and also creating amplifier matching difficulties because of low resistance. The current needed is reduced in proportion to the number of turns, so the formula becomes:

$$I = \frac{\sqrt{A}}{3t} \quad \text{or} \quad \frac{\sqrt{A}}{3.3t}$$

(A in metres or yards) in which t is the number of turns.

It may seem that running a loop of several turns around a hall would not only be somewhat of a chore, but result in a rather unsightly bunch of cables. There is a practical way if avoiding this, that is by using a single run of three-core mains cable, or if the resistance value requires it, two twin-core cables. The cores are connected in series, thereby producing a loop of three turns or four turns respectively.

Cables and resistance

The type of cable used is governed mainly by the resistance required to match the amplifier output impedance and the length of the run. The first step is to measure the perimeter including deviations such as alcoves and doorframes.

Next, determine the resistance needed to match the amplifier output impedance. If a separate amplifier is to be used for the loop, the resistance can be from 5–8 Ω to match its 4 Ω output. If it is to be run from the same amplifier as the loudspeakers, it should be around 7–8 Ω, in which case the loudspeaker system would need to be at least 12 Ω.

With a single amplifier, the loop is fed from the same output as the loudspeakers, so their output levels are linked. The current in the loop is thereby governed by the feedback point of the main public address system.

As the power fed to the loudspeakers is not usually very great, more must be fed into the loop to produce the required current. This can be done by designing the loudspeaker impedance to be higher than the resistance of the loop.

Table 9 Cable resistance

Cores No/Dia (mm)	Area (mm^2)	Current (A)	Resistance (ohms per 100 m)
1/0.2	—	—	57.6
7/0.2	0.22	1.4	8.2
13/0.2	0.4	2.0	4.4
16/0.2	0.5	3.0	3.6
24/0.2	0.75	6.0	2.4
32/0.2	1.0	10.0	1.78
1/0.8	0.5	—	3.6

Amplifier power

Although the production of the magnetic field is entirely due to the current flowing in the loop, voltage must be present across the loop to produce the current, and so power is expended. The next consideration is whether the amplifier has sufficient power to supply this along with its other loads. Combining $W = I^2R$ with the earlier formula we get:

$$W = \left(\frac{\sqrt{A}}{3t}\right)^2 R = \frac{A}{(3t)^2} R \qquad \text{for } A \text{ in m}^2$$

$$W = \frac{A}{10.9t^2} R \qquad \text{for } A \text{ in yd}^2$$

It may appear that while adding extra turns increases the field, it also increases the resistance and thereby reduces the current, so the two factors cancel. However, from the formula it can be seen that the divisor contains the *square* of the number of turns. So increasing the turns reduces the power required.

Let us take an example. A hall dimensions of 12 × 18 m. The perimeter run is thus 60 m plus 12 m extra for doorframes, = 72 m. From Table 9, the resistance of 16/0.2 cable is 3.6 Ω per conductor per 100 m; for 72 m the resistance is therefore 2.6 Ω per conductor. So a twin-turn loop has a resistance of 5.2 Ω.

The power needed is thus 12 × 18/(3 × 2)² × 5.2 = 31.2 W.

For a three-turn loop the resistance is 7.8 Ω. The power required is now 12 × 18/(3 × 3)² × 7.8 = 20.8 W. From this it can be seen that it is an advantage to use more turns if amplifier power is limited.

The three-turn loop using three-core 16/0.2 mains cable will in most cases prove the most practical. However, a smaller hall having a total perimeter of around 50 m has a loop resistance of 5.4 Ω if three turns of 16/0.2 are used. This would be rather low for a loop fed from the same amplifier as the loudspeakers.

The loudspeaker impedance then has to be no less than 20 Ω if the total impedance is to be kept above 4 Ω. Using 13/0.2 cable having a resistance of 4.4 Ω per 100 m, a total of 6.6 Ω can be obtained, which permits a loudspeaker impedance if 16 Ω to match a 4 Ω output.

An alternative is to use two runs of twin cable giving four turns. With 16/0.2, the resulting resistance is 7.2 Ω, which allows a 12 Ω loudspeaker load.

The ratio of loudspeaker impedance to loop resistance gives the inverse proportion of power available to each. So for a three-turn 7.8 Ω loop in an 18 × 12 m hall, requiring 21 W, a loudspeaker load of 12 Ω takes 21 × 7.8/12 = 13.5 W. This can just be met by a 35 W amplifier. The resistance of the loop should always be lower than the impedance of the loudspeaker system so that it receives most current.

Legal aspects and power

A licence is not required for powers up to 10 W but at the time of writing is needed for higher powers. The above examples give around 20 W for a medium-sized hall but this, as has been shown, is a capability to allow for a 10 dB or three times over-average peak. The average actual power in most cases is around 7–8 W for a three-core loop.

For larger halls such as theatres, the average power exceeds the 10 W limit and a licence is needed. It is, however, free, and it is expected that these installations will be de-regulated in the near future and no licence be required.

Separate amplifiers

The matching is straightforward if a separate amplifier is used. A stereo amplifier is quite suitable providing there is a balance control or separate volume controls so that the output of each can be independently adjusted.

The mixer output and loudspeaker amplifier controls need to be set with the latter about halfway up. The loop control can then be set at the required higher level.

However, with the high power outputs now obtainable, and the variety of options for series-parallel connection of the loudspeakers, there is no reason why a single amplifier should not be perfectly satisfactory for most small to medium installations.

Field distribution

So much for the loop's electrical characteristics, now we will consider its magnetic properties. The coils in the hearing aids have a vertical axis the same as the loop and so are in effect a coil within a coil. There is thus a high degree of coupling. If the orientation of the hearing aid changes, as happens, for example, if the wearer bends down, the coupling is reduced and the signal fades.

Field strength within the plane of the loop varies considerably across its width as can be seen from the solid line in Figure 83. From an 0 dB level at the centre it rises steeply be some 20 dB approaching the loop, then plunges to 0 dB at the loop itself to fall to −20 dB beyond it. It then rises again outside the loop, to gradually fall off.

Such a wide variation across its width is not satisfactory. Ideally, a user should be able to sit anywhere, and be able to change position without have to adjust the volume of the hearing aid, but this contour falls very short of that ideal.

There are two reasons for the variations. The first is that the field decreases with distance from the loop, so that starting from the centre, it increases as the loop is approached. The second is due to the shape of the field, which consists of a series of circular lines of force surrounding the cable (Figure 84). At a distance, the lines of force are vertical and so achieve maximum coupling with

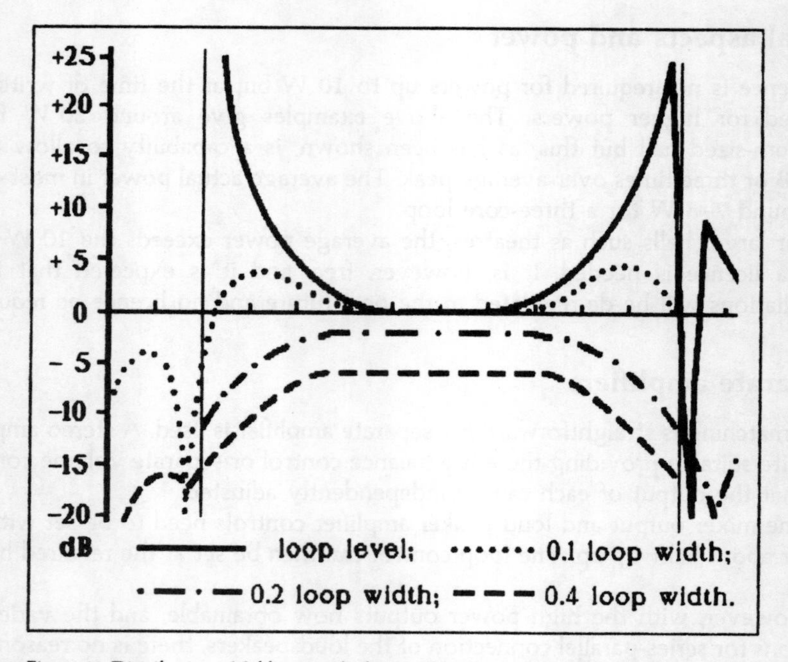

Figure 83 Distribution of field across the loop width at various heights, given as fractions of the loop width.

the hearing-aid coil, but as the cable is approached they curve to the horizontal and the induced signal falls.

When the loop is traversed in a plane that is above or below it by a distance equal to a tenth of its width, the field strength encountered is as shown by the dashed line in Figure 83. The rise as the cable is approached is counteracted by the field curvature, so that it is far less steep and the fall is also less severe near the cable. A slight hump and dip near the cable is the overall result.

At two-tenths the loop width, the response is flat with no humps, but it begins to fall off well before the cable is reached. This is shown by the dotted/dashed line in Figure 83. At four-tenths, the fall-off is still further from the cable, although it is also less steep, as shown by the dotted line. The overall level is well down and the loop needs a higher current to provide the required field strength, (see the power/height table). The response beyond the loop has been omitted on alternate sides in Figure 83 in the interests of clarity.

From this it can be seen that the one-tenth displacement or a little over is the most effective, as it gives a reasonably even field across the width, with a current requirement only a tenth greater and a power a fifth greater than that calculated. These can generally be ignored in practice. For medium-sized halls that are around 33 ft (10 m) in width, the displacement is 3 ft (1 m), which for a seated user, puts the loop at skirting board level. This is a very convenient position for installation.

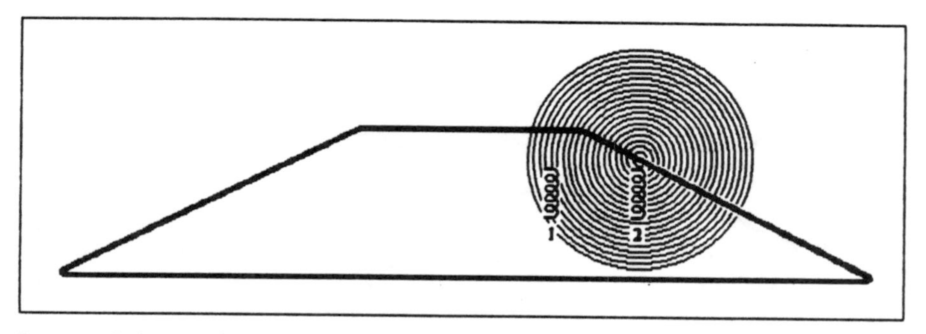

Figure 84 Coil 1 at a distance from the loop encounters vertical lines of flux whereas those at the nearer coil 2 are horizontal, when both coils are not in the same plane as the loop. This acconts for the drop in response near the loop.

Alternatively it can be fitted to the walls above the audience, as its plane can be above or below. For larger halls, this is essential to achieve the one-tenth-width displacement, but for smaller ones the skirting board is the most inconspicuous and convenient position.

Table 10 Ratio of current and power to vertical displacement

Ratio height/width	Multiply current	Multiply power
0.1	1.1	1.2
0.2	1.25	1.6
0.3	1.5	2.25
0.4	2.0	4.0
0.5	2.5	6.25
0.6	3.25	10.6
0.7	4.25	18.0
0.8	5.5	30.2
0.9	7.0	49.0
1.0	8.5	72.2

Where the loop ascends to cross door frames, no adverse effect on the field has been detected, probably because its displacement above the user is much about the same as it is below elsewhere. Outside the loop there is an overspill for an appreciable fraction of the loop width, although at lower levels. Thus adjacent rest rooms and other areas have a useful but limited coverage.

Metal objects such as chair legs seem to have little effect on the results from floor-level loops. Pickup of signal from the loop by microphone cables also seems

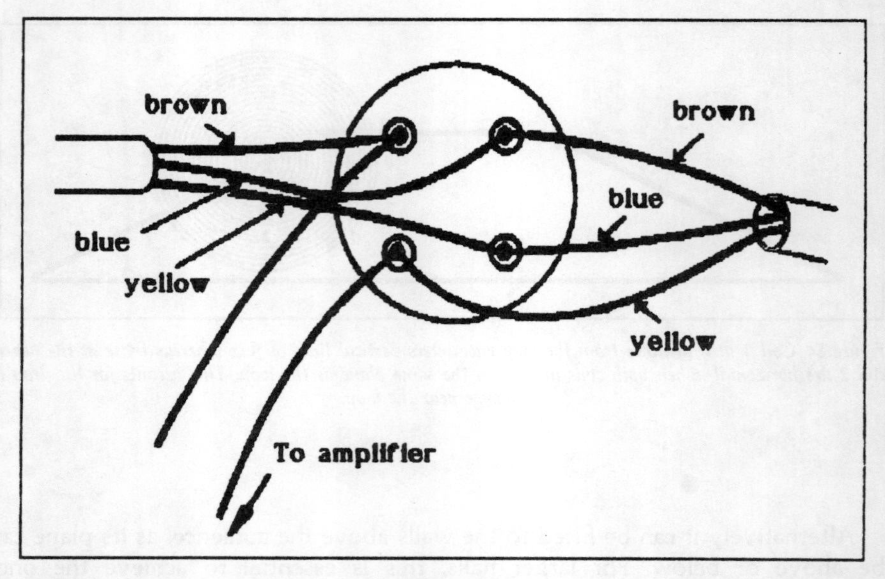

Figure 85 Using three-core mains cable for the loop. Connecting it this way ensures all current runs the same way.

to be negligible and tests at high gain failed to produce any trace of instability with a reel of microphone cable placed within the loop. Once correctly designed and installed, the loop seems to have few problems.

Installation

Installation is accomplished by starting the cable at the point nearest the power amplifier and clipping it around the hall in the designated position. When crossing behind the platform, consider that participants may be wearing a hearing aid and will need to hear what others on the programme are saying. The loop should therefore be raised so that the tenth (or whatever) displacement is obtained with the user in the standing position. This could be achieved by running it at about 3 ft (1 m) from the platform floor, or if this interferes with the decor, it could be diverted to about a 10 ft (3 m) elevation to give a high displacement. This, however, would give very poor results for a seated participant and could also affect users in the front rows, so the 3 ft (1 m) position is much preferred. If desired it could be concealed by burying it in the plaster, but not in metal conduit.

When returning to the starting point, the cable should be cut and both ends connected into a junction box as shown in Figure 85, if using three-core mains cable. Care is needed here because if a turn is connected the wrong way round it will subtract from the field instead of adding to it. From the two free ends a twin cable is connected to the amplifier to complete the installation.

One problem is that potential users unfamiliar with loops are either unaware or forget the need to switch their hearing aids to the 'T' position. It is necessary to remind them to so this otherwise they will obviously not get the benefit of the loop. Notices should be displayed to that effect in the foyer or cloakrooms.

14 Live music

Most public-address systems are intended for speech or recorded music. Live musical performances of a classical nature are usually rendered without recourse to amplification, in fact most concert-goers would shudder at the thought. However, discreet and unobtrusive amplification is sometimes used to reinforce weak-toned instruments such as the harpsichord when played in large auditoria.

Amplification is commonly used for stage, music-hall and cabaret turns as well as for pop groups. With the former, the sources are acoustic, so microphone techniques are involved. With the latter, the instruments are mostly electronic and so are amplified directly.

Vocalists of the pop variety also need amplification as well as electronic enhancement in the form of artificial reverberation, but trained opera or lieder singers rarely use microphones except for very large audiences such as in sports stadiums.

While every effort should be made achieve a clear balanced sound from a live musical performance in a public hall, it is unrealistic to expect results from a public address system comparable to those obtained from a studio recording. Resources are used in the studio that are not available to the public address engineer, and the acoustic environment is under total control. Furthermore there are no feedback problems. However, quite good results are possible by observing a few basic principles, and applying some trial-and-error tests before the actual performance.

The piano

A full concert grand needs no amplification for a solo performance, but does if used to lead community singing. A moving-coil microphone gives good results because its resonance peak produces a bright clear tone, which is necessary for this use. Ribbon microphones tend to emphasize the bass and many pianists tend to do likewise especially when playing for a large audience. Electret microphones also do well here due to their rising treble response.

Feedback is not usually a problem because the sound volume from the piano is sufficient to enable the microphone channel to be operated at low gain. For a concert grand, the microphone should be mounted on a stand possibly with a short boom, and directed on-axis toward the sounding board about 4 ft (1.3m) away. This gives it a low position.

For an upright piano, the microphone should be fitted to a stand with a short boom or goose-neck, behind the piano and pointing downward into its open top. A position just above the mid-range which is just above the middle C is the optimum. Ensure that the microphone is secure otherwise it may disappear into the inside producing some most unmusical effects!

The violin and cello

Sound is produced from the *f* holes on the belly of the instrument and this is where the microphone should be directed. This means a high position some 2 ft (0.6 m) above the instrument, which can be achieved with a stand and long boom set at about 45°. The violin is inclined toward the right-hand side of the player, so the microphone stand should be on his right-hand side. It should not be too close as this could interfere with the bowing.

The instrument does not produce high sound levels, so with its required distance from the microphone, it needs a fairly advanced channel-level setting. Feedback is thus a possibility. A ribbon or electret with a flat response should thus be used to reduce it. Additionally, any microphone with a peaky response can make the instrument sound hard and steely. It has one of the largest numbers of harmonics of any orchestral instrument so emphasis at any region changes the harmonic ratio with dire effect on tone.

Amateur and child players tend to produce scrapes and squeaks, because perfect bowing is very difficult and so is rare among such performers. In such cases, having the microphone low on the left-hand side of the player and directed upward toward the back of the instrument can give a better effect by reducing the pickup of spurious bowing noises.

For the cello the microphone needs to be about $1\frac{1}{2}$ ft ($\frac{1}{2}$ m) from the floor and about the same distance from the instrument. There is less likelihood of feedback because the instrument produces more volume than the violin and so requires a lower channel setting.

The best way of picking up a double-bass is to wrap a microphone in foam plastic and support it in one of the apertures in the bridge of the instrument.

The woodwind

Most of the woodwind family, the clarinet, recorder, oboe, and cor anglais, direct sound downward toward the floor. The microphone should therefore be mounted low and pointing up toward the end of the instrument. A distance of about 2 ft (0.6 m) is about right.

The flute is an exception as it directs sound sideways toward the right of the player. The microphone should therefore be higher and pointing at the end although not in direct line with it. Another position is at right-angles pointing toward the middle of the instrument or even at the player's lips. The latter gives a more breathy effect.

The bassoon propagates its sound upward and to the left of the player, so a high microphone position angled downward similar to the violin is the one to use.

Brass

Brass instruments hardly need amplification in a normal auditorium, but may do in a large outdoor event. The microphone should be directed toward the flare at a distance of about 2 ft (0.6 m) but well off-axis. A brighter tone can be obtained with the microphone on axis but at a distance of 4–5 ft, (1.3–1.5 m)

Choirs and ensembles

These are notoriously difficult to amplify using conventional microphone arrangements. If few microphones are used, the furthest performers are not picked up while those nearby are dominant. The effect can be evened out by placing the microphones at a distance from all, so that the ratio of distance between the furthest and nearest performer to the microphone is small. However, this introduces a high level of reverberation and possibly audience noise, as well as increasing feedback.

The alternative until recently has been multi-milking, in which a large number of microphones are used, each picking up one or a small group of performers. This calls for a lot of microphones, and a large mixer; and getting the balance right can be a nightmare. A good balance achieved at rehearsals may be totally upset by performers moving their chairs a few inches between rehearsal and performance.

Such problems have now all but disappeared with the advent of the boundary or pressure-zone microphone (see Chapter 4). These have an extremely long 'reach' and pick up the back row of the choir equally as well as the front. Furthermore, absence of interference and resulting comb filter effects produces off-axis sound quality equal to the normal.

One or two microphones mounted on acrylic sheets angled downward, suspended at the front and cover the performers will in most cases be all that is needed. The reduced directivity of these microphones may result in greater feedback than the close-milking technique, although it should be no worse than the distant microphone method. Careful experiment and angling of the baffles and the use of a notch filter should produce an acceptable level before feedback.

Any extra care needed with feedback prevention is more than compensated for by the better balance, greater simplicity and better quality obtained with the boundary microphone.

Pop and rock

This is a specialized field in which the performers usually have their own equipment. We will deal here with some of the features.

Each instrument usually has its own amplifier and loudspeaker(s). Guitars are generally of the electric type, although acoustic guitars with magnetic pickup devices attached are sometimes used. The instrument has controls on its body to vary the volume and tone. Foot pedals are also used to obtain special effects such as wah-wah which changes the mid-frequency response and imparts a sobbing effect.

Other effects commonly used include the following: *artificial reverberation, tremulo,* with which an oscillator running at 5–10 Hz modulates the amplitude of the signal, *vibrato,* with which the signal is frequency modulated by an oscillator, *fuzz,* which clips the peaks of the signal to produce harmonic distortion, and *phasing*. The latter is produced by shifting the phase of the signal through a shifting circuit, then combining the result with the original.

Loudspeakers are quite different from those described for normal public address. The drivers are not selected for having a neutral response, but the tone they impart to the reproduction. This is especially so for the lead guitar. The coloration is produced by deliberate distortion from reflections across the cone from the surround, which is obtained by using paper suspension instead of cloth or foam. Other features which research has identified and eliminated for hi-fi units are also purposely included.

Different drive units have different types of distortion and are chosen by the players to suit the sound they want. These should not be used for applications such as vocals where distortion is not desired.

The frequency range of loudspeakers for pop groups need not cover the full range as do hi-fi or public address units. They need cover only the range of the instrument they are designed to reproduce. So, the lead guitar loudspeaker need only have a range of 196–1568 Hz with some extension at the upper end to reproduce harmonics. It thus can be a single 10 inch (25 cm) or 12 inch (30 cm) driver with no tweeter in a cabinet with a non-sealed, acoustically open back. Often two units are mounted in the same cabinet to increase the power handling.

For the bass guitar which has a range of 41–261 Hz, a large 15 inch (38 cm) or 18 inch (45 cm) unit is normal, although loudspeakers having four 12 inch (30 cm) or even 10 inch (25 cm) drivers are sometimes used to give a faster attack along with deep bass. An infinite baffle or bass reflex is needed to produce the lower registers. The I.B. enclosure falls off at a rate of 12 dB per octave below the roll-off point, while the reflex falls at 24 dB per octave. The latter, however, has a lower roll-off. In spite of all this, the response of bass guitar loudspeakers is often to only 50 or even 60 Hz. However, missing fundamentals of low notes are hardly if ever noticed as most of the character and pitch is conveyed by the harmonics.

As the role of the bass instrument is mainly supporting, it does not need a characteristic tone colour. Hence the distorting loudspeakers of the lead guitar are not normally used, most players preferring the smoother tone of a cloth or rubber surround. These are generally less sensitive and so require more power than their paper counterparts.

Vocal

The range of the human singing voice is 73–1046 Hz, but that is from the lowest note of a bass to the top C of a soprano, hardly the range likely to be covered by a pop vocalist!

The limited range can be well covered by a non-sealed cabinet with a 10 inch (25 cm) drive unit similar to the lead guitar but non-distorting. A cloth surround is thereby the most suitable. The amplifier or mixing unit will almost certainly need artificial reverberation to enhance the sound as few if any pop vocalists seem able to produce passable results without it.

While on the subject of the human voice and its frequency range, it may come a surprise to learn that the male voice has a higher frequency content than the female. While the fundamental tones are lower, there are a greater number of high-order harmonics. The female voice has very few harmonics. This accounts for the softer sound of the female and the harder tones of the male.

This has a practical application. Intelligibility in speech is achieved by the production of the consonants, which are mainly composed of transients, hence high frequencies. These transients are therefore more prominent in the average male voice than the female. So for intelligible announcing over a p.a. system, especially in noisy conditions, the male voice is best.

Keyboard

The type of loudspeaker most suitable for keyboard instruments depends a lot on the instrument itself, particularly its range. A large electronic organ may have pedal notes going down to C_4 which has a frequency of 16.3 Hz, but there are many keyboards that have a more modest bass range.

Another factor to consider is the likely size of the audience and auditorium. Very large audiences would require a stack of separate bass, middle and treble units, whereas smaller ones could be well served by one or two full-range loudspeakers.

As all the varied tone colours have been carefully achieved by the design of the internal electronic circuits, no extra modification is required from the loudspeakers as with the lead guitar. In fact, any distortion would detract from the instrument's tonal character.

So the loudspeakers must be up to hi-fi standards. For the smaller audience, one or two full-range units going down to, or near, the lowest note in the range of the instrument, will be needed. The treble range of most keyboard instruments extends an octave or more higher than that of the lead guitar, and many of their stops give effects that are rich in high-order harmonics. The loudspeaker response must therefore go high enough to produce them, which can be achieved with a suitable tweeter.

The tweeter must have a lower frequency limit which overlaps the highest frequency produced by the bass unit. The cross-over is usually 3 kHz. Tweeters that respond down to this frequency often lose a little at the top end but this should not cause undue concern. A response up to 16 kHz is quite adequate as human hearing does not go any higher, and it is unlikely that the majority of the audience can hear even this high if they are frequent attenders at high-powered rock concerts or discos!

A more important specification for a tweeter is its polar response, that is the angle over which it disperses sound. A tweeter diaphragm propagates high frequencies in a narrow conical beam with about a 60° dispersion angle. A rectangular horn fitted to the front extends the side angle to some 90° at the expense of the vertical which is reduced to around 40° depending on the dimensions of the horn. Another type is the bullet tweeter which as its name implies has a bullet-shaped diffuser surrounding a circular horn in front of the diaphragm. This radiates a conical beam over a 90° angle. A wider dispersion can be obtained from a unit fitted with a slot diffractor. This has a horizontal spread of over 120° with a vertical one of 50°.

It is obvious that in a wide auditorium, those sitting at the front sides are at a very wide angle to a loudspeakers standing on stage. They will get no high frequencies at all, and even those nearer the centre will get few unless they are within the range of the tweeter. So horizontal dispersion is an important factor. A wide angle is thus necessary, but as the sound energy is spread out, the tweeter has less 'throw' than when the sound is concentrated into a narrower beam.

High frequencies are attenuated to a greater extent than low frequencies when they pass through air. As they are radiated out from the loudspeakers on the stage, they also undergo greater absorption by the clothing of the audience. So, the sound reaching the back rows will be deficient in the highest harmonics and so will have lost its attack and crispness.

The solution for large audiences is to have an array of tweeters in an arc, mounted in a separate box from the bass unit. This gives a wide dispersion while retaining the long throw of individual units. The spacing and angling of the units can be critical if mutual interference is to be avoided.

Wide vertical dispersion is unnecessary unless a balcony is to be served from the same rig. So for a smaller auditorium, a full-range unit containing a wide-angle tweeter with restricted vertical range could be adequate. For a medium-sized hall, two such units placed either side of the keyboard, angled slightly outward, should give good coverage.

If the bass unit is 15–18 in (38–45 cm) and goes really low it is unlikely to perform very well in the middle frequencies, so a mid-range unit will also be required. However, deep bass can be obtained from 12 in (30 cm) rubber-edged drivers, which also have quite a reasonable mid-response. Rubber surround units need more power to give the same output as those with a cloth edge.

One reason for using a mid-range unit is the Doppler effect at higher powers. A loudspeaker cone moves towards and away from the listener with each cycle. If it does this at a low frequency, while at the same time radiating a higher tone, that high note varies in pitch in sympathy with the lower one due to Doppler effect.

This is unnoticeable at most frequencies at low powers, but when the low note is very low, in the bass region, the cone excursions are large and the variations are slow enough to have an audible effect, especially if the high note is much higher, in the mid or treble range.

So, if the rig is to run at very high power, a three-way system using a mid-range unit, or a nest of them, sounds better than a two-way with just a tweeter and bass driver.

The difficulty is in getting the cross-over circuit right and all the units to match. While an approximate design can be produced for a given set of drivers, it often needs to be fine-tuned by trial and error, and some makers employ computers to do this because of the many variable factors. This is difficult enough with a two-way system, but the problem is considerably compounded with a three-way. For this and other reasons, three-way systems have largely gone out of fashion in hi-fi circles.

So, it all comes down to the power; high power needs a three-way system, but more moderate power is best with two-way systems.

An amount of beaming occurs with mid-range units as well as tweeters. For large rigs then, a pair set at angles in a suitable enclosure gives a wide coverage. If a mid-range unit is fitted in the same cabinet as the bass unit, it should be mounted with an air-tight box around it to shield it from the large air-pressure variations set up by the bass unit.

Cabaret

Cabaret turns often have to perform in restricted venues and to smaller audiences than groups, so the equipment should reflect this. Loudspeakers can be smaller and less bulky to transport. If the turn is only vocal, a single 10 in (25 cm) full-range driver in a non-sealed cabinet should serve the purpose well. Two cabinets may be required to cover a wide area. High power is not usually necessary.

If the stage is low or non-existent, loudspeaker cabinets placed on the floor will be masked by the front row of the audience. There may be no suitable objects on which to stand them, so an essential item of the equipment is a loudspeaker stand for each unit. These are a more substantial version of a microphone stand with fittings at the top to secure the loudspeaker, and are collapsible for transporting. A loudspeaker on such a stand does not excite bass resonances in the platform, which can happen when a loudspeaker unit is supported directly on it. It thus does not produce the often heard bassy, chesty sound, but is clearer and more natural.

Feedback between loudspeaker and microphone can also take place through the flooring from a floor-standing loudspeaker. This is less likely when the loudspeaker is on a stand, especially if the stand has rubber feet. Feedback can and does still take place through air coupling, but the problem is reduced when one possible path is thus eliminated.

If the turn is instrumental, a loudspeaker will have to be chosen that will reproduce the range of the instrument. As sounds made by many instruments have large starting transients, a higher-powered amplifier and loudspeaker is needed than for a purely vocal sound.

In most cases, the best choice is for an infinite baffle sealed box loudspeaker, unless the instrument has a very low register below around 49 Hz (G_3) in which case a reflex system would be better.

Monitors

When a group is playing behind a stack, it can be difficult to hear just what the effect is, especially if the audience is noisy. A monitor or 'fold back' loudspeaker is thus necessary to feed sound on stage. This needs to be a full-range unit although extreme bass and treble are not necessary. High power is not essential, and could lead to feedback with the on-stage microphone if the volume is too high.

The choice of cabinet type is therefore not critical. An infinite baffle or reflex can be used, but the former is the most straightforward. A two-way system consisting of bass unit and tweeter is suitable, but for less complication and an uncoloured result, a twin-cone full-range driver is capable of excellent results.

A wedge shape is much favoured for the cabinet with the front sloping upward. It can thus be placed on the floor of the stage, and be directed toward the head of the performers. A refinement which can be useful is a volume control enables the level to be adjusted during performance if need be without upsetting the balance of other units that may be operating from the same amplifier. Special high-power controls of 30 W rating are obtainable for the purpose.

In the case of performers with acoustic instruments, especially small ensembles playing in acoustically dead surroundings such as a stage heavy with curtains or in the open air, monitoring is essential. String and woodwind players in particular tend to 'force' their tone if they cannot hear their instrument very well.

What is needed in these cases is a simulation of a good studio acoustic with natural reflections. In television studios this is often provided by an ambiophony system. Loudspeakers arranged at various places around the performers, but out of sight, are fed with signals having various degrees of delay. These are derived from the main audio signal but fed through delay devices such as electronic delay lines or endless-loop tape machines.

A simpler set-up is to use artificial reverberation. A reverberation unit is fed from the main signal and supplies a separate slave amplifier. This feeds several loudspeakers around the performers. The floor wedge-type cabinets are ideal as they are inconspicuous as well as directing sound toward the players. As it is only reverberation that is reproduced, the drivers can be small and basic.

With either system, feedback is a possibility as the loudspeakers are directed toward the microphones, so gain must be kept well below the critical level.

15 Outdoor public address

Indoors or out, the requirements of a good public-address system are the same. There are a number of special considerations, though, that need to be considered for the outdoor system.

The loudhailer

The loudhailer is the simplest and most compact form of outdoor public address system. It contains a microphone, loudspeaker, amplifier and power supply all in one hand-held unit. The loudspeaker is a re-entrant horn which affords the high efficiency necessary so that maximum sound output can be obtained from a small-capacity power supply, usually torch batteries. The disadvantage is a very restricted bass response, and the device is very directional.

The microphone is a high-output type and is spoken into at a range of a few inches. The amplifier thus does not need a high gain, and this helps to conserve battery power. Further economy is achieved by a trigger on/off control which is depressed only when actually speaking. The amplifier output stage is biased to class B to reduce standing current to a minimum; the distortion thereby created is insignificant for the intended use.

The amplifier is contained in a cylindrical extension of the horn, and the microphone is mounted at the opposite end so that the user appears to speak through the instrument like a megaphone. A pistol grip provides the support and also houses the batteries, with the trigger (Figure 86).

Some of the larger ones have a shoulder strap or short rear legs for support in which case the microphone is separate, connected by a curly lead. Noise-cancelling microphones are common with these models.

Power is around 6 W for the smaller ones and 10 W for the larger. Under favourable conditions the large ones are audible for up to a mile, but a quarter of a mile is obtainable in most situations.

Simple portable systems

Other portable systems take on a more conventional form. Consisting of a single microphone on a stand with an amplifier, batteries and loudspeaker, the problem is how to make it all pack away into a convenient form for easy transportation.

Various arrangements have been used. It is common for the amplifier and batteries to be housed in the loudspeaker case which also has a compartment for the collapsible microphone stand. Other systems have folding or collapsible columns and an amplifier that forms a solid base for the microphone stand.

The main problem with all of these is the power supply because, unlike the loudhailer, the amplifier is on continuously. Also the system is usually used for

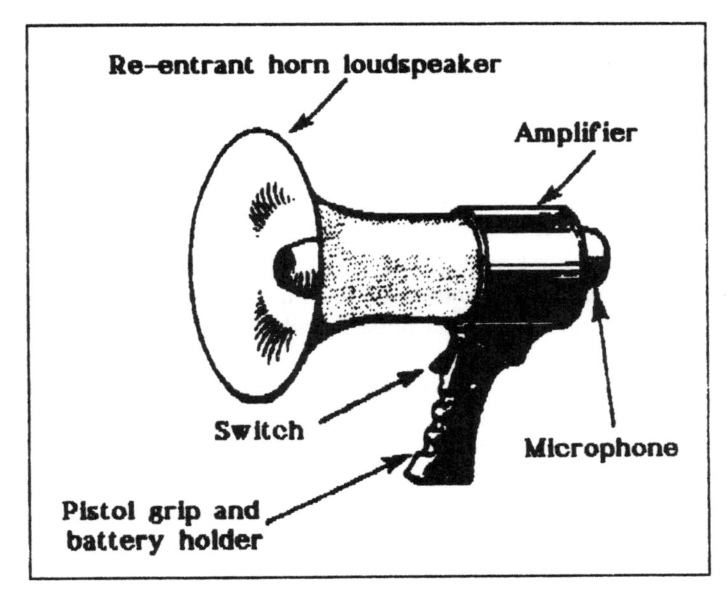

Figure 86 Loudhailer.

speeches, presentations and the like whereas the loudhailer is generally used for short spot announcements. Larger batteries are therefore required, preferably rechargeable ones.

One compact system consists of a leather shoulder bag containing everything including a conventional loudspeaker. The microphone, which is connected by a curly lead, is hand held and stowed in the bag when not in use. The bag can be stood on some support with the loudspeaker facing the audience or used from the shoulder. Range is not great because the loudspeaker is not a horn, but is adequate for addressing a small group.

The sound car

These are generally used for spot advertising announcements or public service information while travelling slowly, or for outdoor public meetings when stationary. The loudspeakers are roof-mounted horns, usually re-entrant and pointing fore and aft.

Power ratings vary from 8 W to 40 W, and the lower frequency response from around 450 Hz down to 200 Hz. Many horn loudspeakers have transformers for 100 V line working, but these are not necessary for sound car use.

Although a good bass response is not necessary for speech, especially for short announcements, reproduction that is too deficient in lower frequencies can be irritating and sound tinny and 'cheap', not the sort of image an advertiser would want to convey. Furthermore, while some bass loss can improve intelligibility, too much can degrade it by emphasizing sibilants. Horn loudspeakers should therefore be chosen that have as good a bass response as possible for the size.

The amplifier is supplied from the car battery and so can be of a higher power rating than the loudhailer or other portable systems. From 25 to 40 W is usual. Some car amplifiers have built-in siren and other effects. Another option is an integral cassette player to broadcast music or even the announcements, so leaving the driver totally free to concentrate on driving. For a purpose-used sound car, the amplifier is installed in the dashboard with a clip for the microphone and wiring run behind the trim.

Microphones are hand-held with a spring-loaded pressure switch lever on the left-hand side. A slight squeeze switches the unit on. Some have anti-phase dual transducers for noise cancelling.

It should be noted that local bye-laws may restrict the use of sound cars or any amplifying equipment in public places. The details may differ from one authority to another, so it is advisable to enquire just what the position is for your area.

Outdoor events

Many outdoor events of all types and sizes require the use of public address equipment. Some may be temporary, others permanent. Just what is needed depends on the nature and size of the event.

One problem with indoor systems that will not normally be encountered outdoors is feedback. There are few if any hard reflective surfaces, and the sound is propagated upward and away from the microphones with normal air-temperature gradients. The various measures for defeating feedback described in Chapter 11 will therefore be unnecessary.

This changes the picture with regard to choice of microphones. Moving-coil units are the most suitable as these are robust and can stand the hard usage and knocks often received with outdoor use. Omnidirectional types are also best as they are less prone to wind noise than velocity directional models. Wind noise, in fact, takes over from feedback as one of the major problems.

A wind shield should always be used even on a calm day, as an unexpected breeze can suddenly spring up. The sponge-foam pop shields often supplied with microphones help to a limited extent, but air turbulence around the shield can still generate noise. This can be reduced by fitting a bag over the foam which is made from a soft loose woven material. Draw strings at the opening of the bag can be tied to ensure that it does not slip off or blow away. Suitable bags could be knitted with coarse wool on large needles.

Rain is always a possibility, in fact many outdoor event organizers with long memories of past events would probably say it was inevitable! Providing the front and side microphone vents are well covered with a foam wind shield there should be no cause for concern. However, the shield should be taken off as soon as possible afterward and dried out, and any moisture that had penetrated wiped off. It should be noted that fitting plastic bags or skirts over suspended microphones modifies the response, giving a hard unpleasant reproduction.

Loudspeakers

Fewer loudspeakers are usually required for a given area than are needed indoors, because without feedback problems, they can be driven harder and so cover a wider area. The lack of reverberation (except with stadiums) gives greater intelligibility, but this advantage can be reduced by ambient noise and wind.

The type of loudspeaker used depends on the application. For announcements, the horn is quite suitable as it can give high sound levels. A nest of horns on an elevated mounting, facing in different directions can cover a wide area. Although an extended bass response is not necessary or even desirable for speech, the hard tinny sound of small horns is unacceptable even for announcements. Where used, the larger models having a reasonable response in the bass should be employed. There is usually no space problem mounting them outdoors, but transporting them for temporary set-ups may create difficulties.

For public meetings or other events involving lengthy periods of speech even the large horns with their pronounced mid-frequency response, can be irritating to listen to, especially at high volume. The line-source or column loudspeaker is the preferred choice here. Having an attenuation of only 3 dB for each doubling of distance, they have a long 'reach' and can cover considerable distances. They are suitable for both permanent installation and are easily transportable for temporary systems. However, their limited vertical sound propagation, while making them eminently suitable in some situations, makes them less so for others.

A factor which now must also be taken into account is vandalism. Loudspeakers mounted within easy reach will almost certainly become a target. Fortunately, the best position for good coverage is usually an elevated one, which is also the least accessible.

Prevailing wind and rain

The weather is as always a major consideration. All loudspeakers should be weatherproof. This is no problem with a permanent installation as suitable models are available. For a temporary system, the units may also be used for indoor set-ups, and so lack the necessary weatherproofing.

The most convenient way of keeping out rain is by encasing each loudspeaker in polythene lie-flat tubing. This is available from the larger plastics suppliers, and consists of rolls of flat clear polythene seamless tubing of various diameters that can be cut to length.

The tubing should be cut a little longer than the column length, so that it can be turned over at the top and sealed with sellotape. The bottom should be left open so that any condensation can run out. This is much easier and quicker than wrapping columns in plastic sheeting. While this is usually discarded afterward, the tubes are reusable and can be rolled up and stored for a future use. Unlike the microphone, sound quality seems to be unimpaired by encasing the loudspeaker in plastic.

With stadiums, the location of the loudspeakers is dictated by the layout of the stadium itself. For events such as school sports days, gymkhanas and the like where spectators are not accommodated in any particular structure but are free to move around the ground, loudspeakers must be strategically placed to achieve the best coverage.

While winds can come from any direction, for a large part of Britain, the prevailing winds are from the south-west. Sound from loudspeakers placed at the south and western boundaries will thus most likely be wind-assisted to reach a good distance over the required area.

Getting a forecast of the likely local wind direction before the event is a good idea just in case it should differ from the usual; then the loudspeakers can be positioned to take advantage of it. In addition a nest of loudspeakers facing in all directions could be installed near the centre of the area. These would reinforce those at the boundaries, and give some coverage should the wind change.

Auxiliary loudspeakers

These may be needed in hospitality tents, cafeteria marquees, information booths and so on. Single cabinet loudspeakers will suffice in most cases, although marquees may need one or more columns.

It is wise to run all auxiliary loudspeakers from a separate feeder from those serving the main area. Auxiliary feeders are vulnerable to all kinds of damage and any fault, especially a short-circuit, could otherwise silence the whole system.

Loudspeaker feeders

As long runs are involved in most cases, 100 V line working is necessary. Feeders are usually run at a high level above public reach, and any nearby points such as outbuildings, flag poles, marquee posts, and trees can be used to support them. When using trees, though, beware of branch movements causing wind which would break the cable.

There is a limit to the amount of unsupported feeder length that can be suspended between two points. Twin 16/.02 feeder will support a maximum of 3 lb (1.362 kg) dead weight, and double that, 6 lb when supported at two points. The cable weight about 1 lb (0.45 kg) per 12 yd (approx 11 m), so $12 \times 6 = 72$ yd (66 m) is the maximum length.

This assumes that the supporting points are smooth and round, thus distributing the load over a few inches instead of at one point; also that the run is truly horizontal. If it is not, the higher point bears proportionately more of the strain than the lower one. A further factor is wind, which will add to the strain, especially if gusty. So, the practical maximum is 40–50 yds (36–45 m). Thicker cables such as 24/.02 support more weight but themselves are heavier, and nothing is gained by using them. A supporting catenary wire is the best option.

Chuting down-pipes can make a convenient anchoring and support point for a feeder, but do not pass the cable around the pipe. The small gap between the pipe and the wall may make recovery difficult if the cable is being wound onto a reel. Instead, loop the cable and tie it as shown in Figure 87.

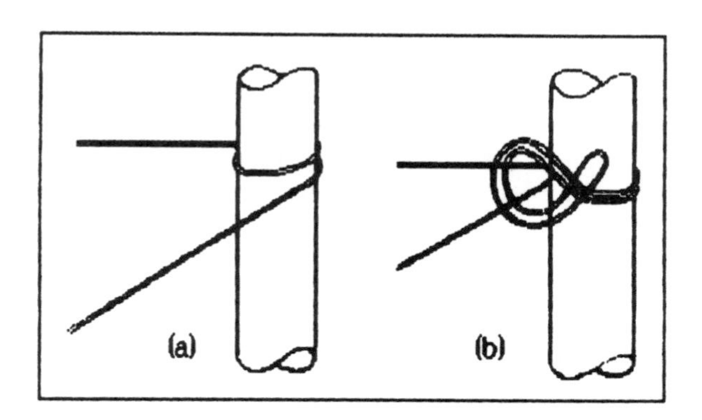

Figure 87 (a) Feeders passed around pipes may be difficul to recover later; (b) they should be looped as shown.

There are other hazards for overhead feeders, such as people carrying ladders and high-sided vehicles. Both have resulted in cables being brought down. Adequate height is the only precaution. BS 6259 stipulates 3 m minimum.

An alternative method is that of burying the feeder. This is not as inconvenient as it sounds. A sharp half-moon lawn edging tool can quickly make a groove through turf a couple of inches deep. With a helper following behind laying in the feeder and treading it down, quite a long run can soon be completed. Not, however, a recommended method across bowling greens!

Recovery can usually be easily made by just pulling up the cable as it is wound on the reel, treading down the disturbed turf afterwards. The main snag is that the cable will most likely be rather muddy.

One important point with buried cable is that there should be no pinholes from previous loudspeaker connections or other bare parts or taped joins in the insulation. Any such would cause an earth leak which could produce instability.

Power supplies

A mains feed can often be obtained from the nearest supply, but the greatest caution is required in handling it. Touching a live lead indoors when standing on a carpet or dry wooden floor produces hardly a tickle, but when standing on concrete or damp earth, it most likely would be fatal. The reason is that the subject is in good contact with earth and so is effectively connected directly across the supply, the 'neutral' side of which is earthed.

Ideally, the mains supply should be fed via a large isolating transformer at the supply end. Neither end of the secondary is earthed so that accidental contact with either conductor has no effect. If a transformer is not available, the area where the amplifiers are operated should have some insulated flooring, if only a large plastic sheet or tarpaulin.

A good earth should be provided, preferably on site rather than the mains earth which may involve a long cable run. There is no harm in having a local earth as well as the mains earth as a back-up providing they are joined together at the system distribution point. Do not earth one part of the system to the mains earth and another to the local one as this would produce an earth loop and almost certain hum. A residual current device (RCD) should be used in the supply, preferably not a plug-in type as these can be removed or by-passed. It should be rated at 30 mA.

For really remote sites far from mains supplies the choice is between a petrol generator supplying mains voltage, or car battery operation with 12 V amplifiers. The wattage rating of the latter is not a steady consumption but that taken at maximum peaks. So, a 40 W amplifier may average no more than 24 W which at 12 V corresponds to a steady 2 A drain. A fully charged 60 ampere-hour battery should thus last 30 hours; if two amplifiers are used, 15 hours or a little less. However, a spare fully charged battery should be on site.

As with indoor systems, ensure that all the loudspeaker transformer tappings do not exceed the power rating of the amplifier. When operating small battery amplifiers, it is easy to overlook this, especially as volume levels, hence power tappings, are usually higher outdoors. It needs only four 10 W columns to fully load a 40 W amplifier.

For moderate-sized set-ups, a couple of battery amplifiers with car battery is the best arrangement, but for large systems, mains amplifiers with a petrol generator of sufficient output is the most practical. In such a case, install the generator on the leeward side of the site so that the noise and fumes are blown away, not over it.

Sports stadiums

The type of system rather depends on its purpose and permanence. The sports days/garden fete type of installation previously discussed are of a temporary nature, set up for one or two days only. Since the Hillsborough football stadium disaster, public authorities have tightened up on all safety aspects of sports grounds. One feature to receive attention is that of the public address system. All stadiums must have a permanent system which must be able to broadcast clear and audible instructions for crowd information and control in the event of trouble.

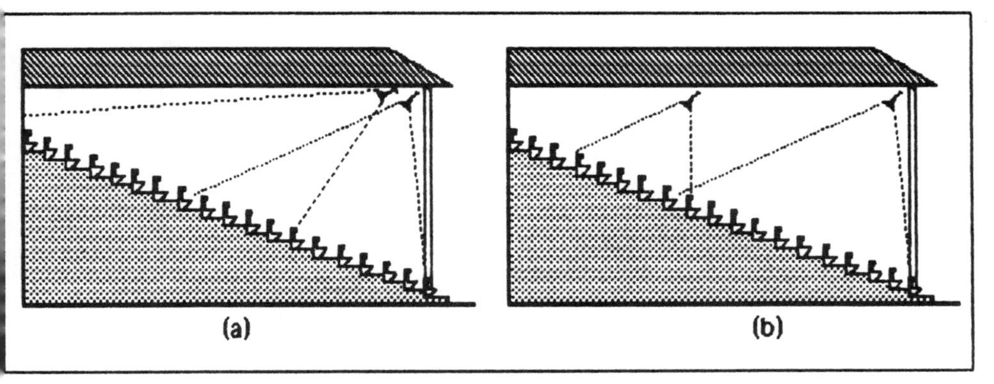

Figure 88 (a) Two rows of horns angled to cover the stadium; (b) if the second row is moved further back its range may be insufficient to cover the back rows.

In general, loudspeakers should be mounted high and aimed downward at the opposite boundary as recommended in BS 6259. This gives coverage of the intervening area and also rapid sound level fall-off beyond the boundary, so avoiding annoyance to those living nearby.

A problem with this type of system is echo and reverberation, common in grandstands due to highly reflective surfaces, high sound levels required, and the distances which can produce long delays between reflections. A further problem may arise with smaller grounds where grandstands face each other across the pitch. Delayed sound from the one can reach the other, adding to its own echoes and reverberation.

To reduce these effects, reflections should be minimized by directing the sound into the spectators and not into the walls and particularly the roof of the stand. Furthermore, loudspeakers should be located so that no part of the stand received sound from widely separated sources. How that is achieved depends on the layout and construction of the stand.

One method which satisfies these conditions is to position horns in a row along the front underside edge of the canopy. If they are angled so as to cover the front rows, they are unlikely to reach to the back, so another row angled higher may be needed to serve that area (Figure 88(a)).

It may appear that the second row would be better placed nearer the back, but its range would then be much less than the first because of being nearer the seats (Figure 88(b)), and might be insufficient to reach the back rows. Yet the power level would have to be lower to avoid deafening those underneath it.

Another possibility is to mount re-entrant horns at the back wall so as to project sound at grazing incidence over the seats. A single row of loudspeakers would then be sufficient, but the sound level near the horns would be unbearably high so as to give sufficient volume to reach the front. No part of any spectator area should be very close to horn units.

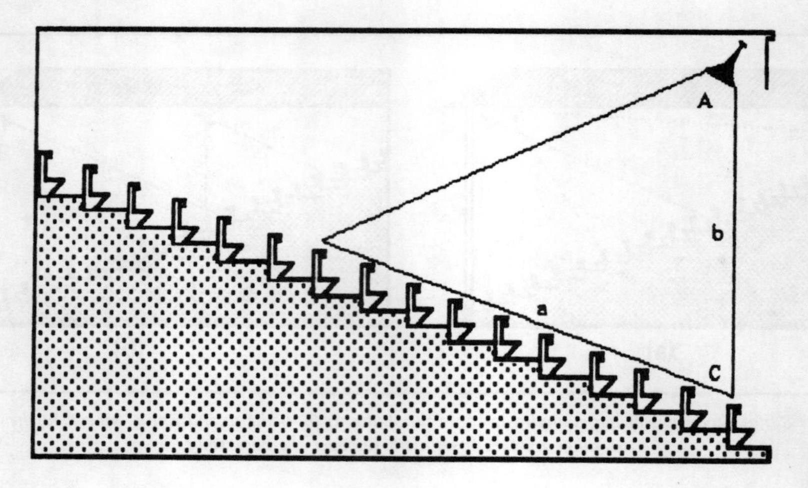

Figure 89 Coverage of horn can be calculated from

$$a = \frac{b \sin A}{\sin (180 - A - C)}$$

Calculating the range

Looking at Figure 89 the angles and indeed the necessity for a second row can be worked out from the dimensions of the stand and the dispersion angle of the chosen horns given by the makers. We need to know the height of the canopy above the first row (*b*), the angle of dispersion of the horn as given by the makers (*A*), and the angle of the seating from the perpendicular (*C*). If the makers do not give the dispersion angle it can be assumed to be a minimum of 60°, but it is best to get this information if possible as it could save installing unnecessary units.

Using basic trigonometry:

$$a = \frac{b \sin A}{\sin (180 - A - C)} = \frac{b \sin A}{\sin (A + C)}$$

a then gives the range of the horn from the front row in the same units as used for the height of the canopy. As the seat rows are usually equally spaced it is then more convenient to translate this distance into rows of seats for installing the rest of the horns. They should be mounted and aimed at the middle row, and placed at distance *a* apart. The number required can be readily calculated from this.

With shallow stands and horns of wide dispersion angle the range may cover right to the back or nearly so, in which case the single row will suffice, If there is a large short-fall, a second row will be required.

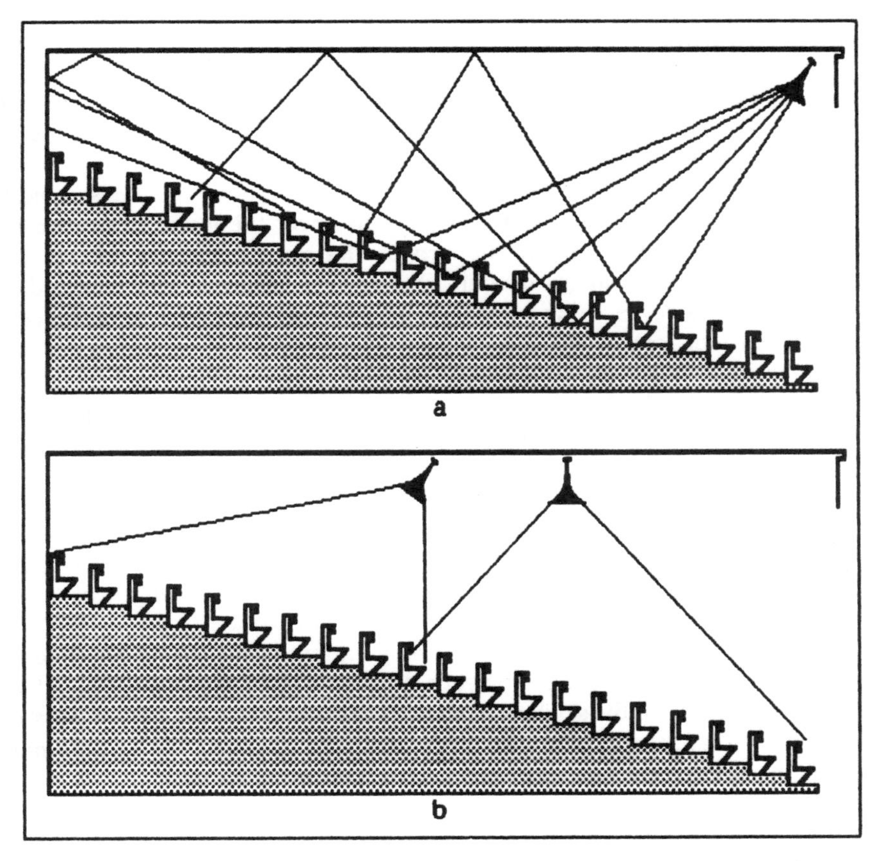

Figure 90 (a) A deep stand with a shallow rake may produce multiple reflections from the front horn at the rear; (b) this can be reduced by angling the front horn downward and moving it further back.

If the stand is deep from front to back, yet the rake of the seating is shallow, this arrangement may result in poor intelligibility at the rear due to the front row of horns producing multiple reflections from the floor, canopy and back wall (Figure 90(a)). In this case sound from the front horns should be directed so as not to be reflected toward the back from the floor, but aimed more directly downward.

This can be done by moving them inward from the canopy edge as shown in Figure 90(b). The range of coverage is thereby altered as the dimension b is changed, but not to a large extent. It can be calculated by taking the square of the canopy height b plus the square of the distance of the row from the canopy edge d, and finding the square root of their sum. This will give the value e which can then be used in place of b in the above formula, or $e = \sqrt{(b^2 + d^2)}$.

Using columns

Columns may be considered in these positions and angles for their superior sound quality, but although their coverage would be wide, their limited vertical propagation would serve only a few rows.

If there are a number of vertical supports for the canopy, columns could be mounted on these, one mounted fairly low to cover the lower rows, and one higher to reach the back. If the supports are far apart it may be necessary to have a nest of three columns in each position, one facing to the back and the others to each side. However, modern grandstands tend to use cantilevered canopies without frontal supports.

Another possibility for using columns is in a row along the back angled downward. Unlike the horn, the column does not produce a very high sound level nearby to the distress of spectators there. The main consideration in this case though would be that of possible vandalism, as in most cases the columns would be within reach.

For seating underneath the balcony there are similar problems, but these are compounded by the roof being usually much lower. The spread and range of any unit mounted under it is therefore much less. Also the length of a horn with good bass response may be difficult to accommodate.

A solution to the size problem is to use re-entrant horns. Adequate coverage can be obtained by having a pair in each position angled sideways at about 45° as well as downward. By similar calculation to those given above the pairs can be spaced so that there is a certain overlap with the next pair, but not too much so that spectators near one can hear much sound from the next.

Open terraces

These are comparatively easy to serve as there is no reverberation or echo, other than sound heard from the grandstands. A couple of towers with a nest of loudspeakers mounted high and aimed to cover all parts, one each side of the terrace is probably as good an arrangement as any. Columns can be used here to advantage, but some must be tilted forward, do not rely on high placed columns aimed at a distant point to cover nearby areas. The towers used for floodlighting on some grounds may prove suitable for supporting these.

Wiring

As the public-address system is an essential emergency facility, its installation must be done in such a manner that it is not immediately jeopardized by any disaster situation. For example a fire starting in a part of a grandstand that carried its loudspeaker feeders could quickly destroy them and silence the system for the whole stand. Physical impact at a critical point could sever cables.

To reduce such possibilities, main feeders should always be run in conduit. Conduit should also be used for the loudspeaker rows with outlet boxes to serve each unit. Cable from the box to the loudspeaker need not be so protected as only one unit is involved.

An additional safeguard is to run two feeders in the conduit, each supplying alternate loudspeakers. If the feeders are brought to the middle of the stand they can then divide and supply two legs, one going in each direction. Each leg can be of a single run or can consist of two cables feeding alternate units. The twin main feeders can be fed from two separate amplifiers or commoned and supplied from a single high-power amplifier. The former ensures some sound in the stand should an amplifier break down at a crucial moment.

When crossing open spaces it is best for the feeder to be buried in conduit. This actually may be more convenient than running it around surface structures. Free suspension across an open space is risky and should not be used for a permanent installation. When crossing from one stand in which the amplifying equipment is housed, to the opposite one, it would be convenient to do so by burying feeders across the pitch— with the co-operation of the groundsman, of course!

Amplifiers

As the system is to be used only for announcements and possibly recorded music, an elaborate mixer is not required, in fact a combined amplifier–mixer having two or three inputs will suffice. Other slave amplifiers can be fed from the line-output of the first to feed various parts of the ground, and each should be clearly and permanently labelled to show which areas it covers. Thus announcements can be channelled to specific areas if required and not to others. The whole system should be rack mounted and interconnected inside the rack to avoid interference and tampering.

All this would be of little use if the mains failed, so it is wise to have a separate mains circuit supplying the amplifiers alone and separately fused.

The arrangements we have described may seem to be very much a case of overkill for merely making announcements and playing a little music. Hitherto, most stadium public-address systems have been very rudimentary. However, it must be borne in mind the possible situations for which it may be needed. It could reduce the effect of a major disaster. Apart from this, many local authorities are now demanding effective systems as a condition of granting permission to operate.

Temporary stadium systems

Temporary systems are quite different from permanent ones in purpose, quality and layout. They are usually required for some special event such as a large convention, meeting or rally, for which indoor premises would be too small.

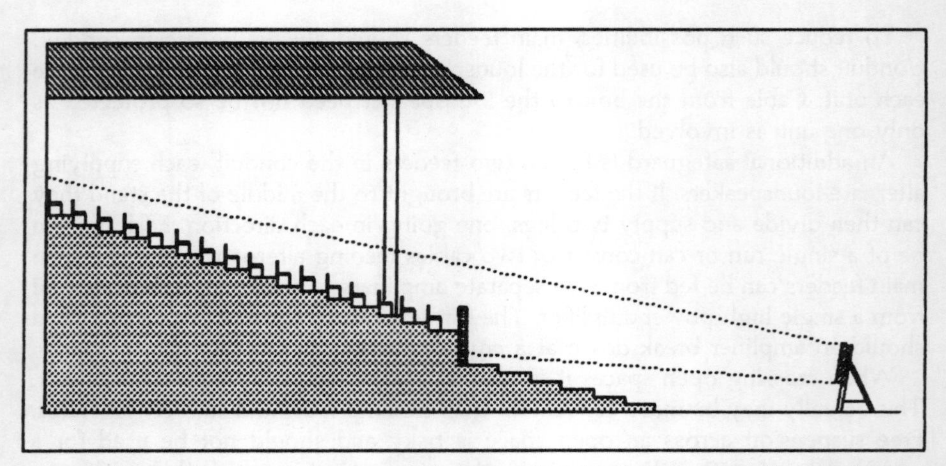

Figure 91 By careful angling of the column minimum sound reaches the canopy to produce reverberation.

The control and mixing facilities of the permanent system would in most cases be inadequate, as several microphone and other programme sources are likely to be needed for the special event.

As lengthy programmes are involved, the sound quality of the permanent system, of which horns probably make up a major part, is unlikely to be good enough. Installing columns in the grandstand canopy or on towers for better quality would be a major task and would not give the desired coverage as we have already seen.

Fortunately, the most effective system is also one that is simple and easy to install. It consists of column loudspeakers mounted on easels around the perimeter of the pitch at intervals of about 10–12 yards (9–11 m). These are directed at an angle up into the stadium.

Columns normally have a wide horizontal propagation pattern (see Figures 62 and 63), but the pressure wave from any column within the row is prevented from spreading sideways to any great extent by the pressure wave of its neighbours. So it is is contained within a fairly narrow corridor in front. The sound energy is thus concentrated and so is maintained over a very extended range which can reach to the back of the deepest stadium. This concentration is aided by the contour of the stadium which narrows toward the back because of the rising floor.

Pressure waves from the last column at the end of a row are not concentrated like the others because there is no neighbouring column to exert side pressure. So its pressure spreads sideways, its range is reduced and sound levels fall. To minimize this effect the last couple of columns at each end of a row should be tapped to a higher power rating, or alternatively spaced closer together than the rest.

Figure 92 Using a sighting tube to angle a column.

The individual pressure waves from all columns build up into single in-phase flat-fronted pressure waves at some distance from the row similar to the indoor LISCA system (see Figure 71). Thus the echo effect usual with multiple sources is avoided. Resulting intelligibility is very good.

It should be noted that this is exactly achieved only if the row is in perfect line. Any units out of line could result in cancellation comb filter effects at short wavelengths.

The flat-topped propagation characteristic of column loudspeakers is exploited by carefully angling the columns so that the minimum sound is directed into the canopy, so reducing reverberation (Figure 91). Angling can be done by means of a sighting tube laid across the top of the column and aimed at the back row of the stadium (Figure 92).

When there is a balcony, this can be covered by another row of columns angled into it (Figure 93). These need not actually be a separate row but can be alternated in a single line with those serving the lower deck.

Feeders can be laid out along the row and Joyce-pinned connections made to all except the last column, to which the end of the feeder will be directly connected (see Appendix). It would be wise to support connections off the ground by tying each one to its easel, otherwise rain could produce earth leaks. Quick-release knots will facilitate rapid recovery after the event. Two feeders should be run for each row, each supplying alternate columns. Any fault on one would then leave half the loudspeakers still operating.

It is a good practice to operate a microphone mixer on or near the platform or whatever is being used to conduct the proceedings. The operator can then see what is happening, and only a single high-level feed is required back to the amplifiers, instead of a number of low-level microphone cables.

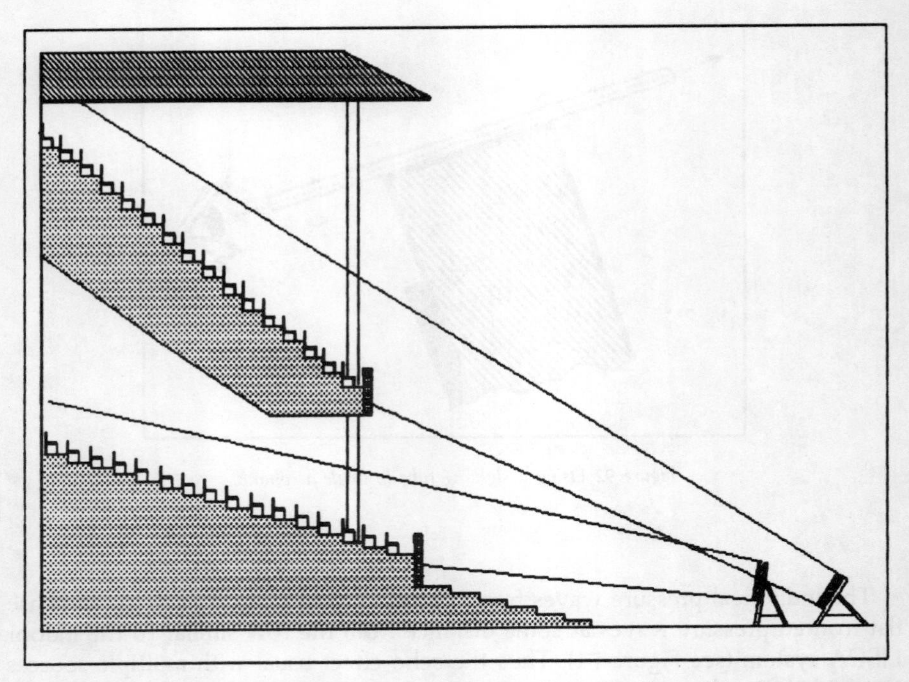

Figure 93 A balcony can be covered by a second row of columns suitably angled. These need not be staggered as shown from the first.

16 Special systems

In this chapter we will take a look at a few special situations not covered previously. The general principles already discussed apply to these, but with some modifications and additional considerations.

Factory systems

The requirements for factory systems are quite basic, to broadcast announcements, relay music, produce time and in some cases alarm signals. Conditions differ considerably throughout the factory, some areas are large, others small, some noisy, others comparatively quiet. The loudspeaker system installed for each should not be the same for the sake of uniformity, but should be designed for the particular situation.

Take the machine shop for example: here there is a high level of noise and operators should be wearing ear muffs for hearing protection. The maximum exposure levels for given time periods are shown in Table 11.

Table 11 Maximum exposure levels

Sound level (dBA)	Time	Sound level (dBA)	Time
90	8 hrs	102	30 min
93	4 hrs	105	15 min
96	2 hrs	108	7.50 min
99	1 hr	111	3.75 min

To be intelligible, sound levels for speech should be at least 10 dB above the ambient noise. Here, such a level would be impractically high, and in any case would be rendered inaudible by the hearing protection.

The solution to this problem is to install an induction loop. In place of ear muffs, operators can be issued with headphones containing an induction pick-up and amplifier as well as provide insulation from outside sound. Industrial models are available for this purpose.

It may also be necessary to install some loudspeakers for the benefit of any personnel who may have temporary business in the area and so not be wearing hearing protection. Although they may not hear announcements over the noise, alarm and time signals swould probably be heard.

In some cases, it may be required to cover large areas of less noisy machining operations. Sound from distant loudspeakers may be obscured by ambient noise, so overhead units are probably the best solution, either mounted in the ceiling, or if the ceiling is high, suspended below it. Depending on the noise and area occupied by each machine with its operator, one loudspeaker could be allocated to two or four work stations.

In some areas the workers are mobile and noise comparatively low. Stores, paint shops and loading bays are possible examples. Long-range units with good coverage are needed, and a nest of columns for large areas or one or two for smaller ones would serve well here.

For offices, ordinary cabinet loudspeakers will be quite suitable unless the area is large, in which case a column would give the required range without being too loud for those nearby.

The distribution system should be 100 V with transformers at each loudspeaker so enabling individual power levels to set. Wiring should conform to the current factory wiring regulations.

The number and the power of the amplifiers is determined by the size of the premises. Some means of routing signals to various areas should be incorporated, either by having separate amplifiers, or by switching loudspeakers at the control point.

An automatic-reverse cassette deck could be the music source. The microphone input should have an auto override so that the music volume drops or is cut completely when the microphone is used. Various time and alarm generator modules are available and suitable ones can be chosen for the required purpose. All amplifier and input modules should be rack mounted.

Council and debating chambers

The special feature here is the large number of participants who may be spread out over the area with tables or desks. In addition, there may be a further area such as a gallery to which the public are admitted.

A large number of microphones are required with mixing facilities, and feedback is an obvious hazard. The easiest part of the job is the public gallery in which a pair of columns can be installed and angled to give good coverage and reduce sound reflection to the chamber.

Loudspeakers are also needed within the chamber so that all participants can hear the proceedings. Here again columns are the best as these can be so positioned to minimize direct pickup by the microphones.

If tables are used, the best solution is to use boundary microphones let into the table tops, as one unit will serve several participants and so reduce the number needed. Then the common problem of speakers not speaking into the microphone will not arise.

If tables are not used, boundary microphones may be suspended overhead on small transparant acrylic baffles angled downward Experimentation is needed to find the best positions to give maximum pickup without feedback.

Feedback is reduced if only one microphone is live at any time, so some form of control should be available so that the chairman can select the appropriate microphone when calling on someone to speak. This will also reduce the effect of interruptions! However, fiddling with faders to get the right level without feedback is not likely to be popular with chairmen who are trying to concentrate on the debate.

It is better to install the mixer elsewhere with all faders pre-set well below feedback level. The chairman's control panel then consists of a row of labelled switches which merely turn each microphone on or off. All are normally off, and the appropriate one switched on when required. The switches should be wired to short-circuit the output of each microphone when in the off position to avoid hum.

Such a system does not provide high volume levels, but as the chamber is most likely of moderate size, the system can be regarded as one giving sound reinforcement, to augment the speaker's natural voice.

Overflow meetings

Running loudspeaker feeders to an overflow hall was dealt with in Chapter 10 and is only practical for short distances. Distant venues can be linked by means of a British Telecom line if available. As some time is needed to install a line at each venue, plenty of notice should be given. Matters are simplified if there is already a line at one or both of the venues.

The line cannot be used as a loudspeaker feeder as BT allow only 1 mW maximum to be fed down the line because of the possibility of interference with other lines. Some means of limiting the output to the line to below this must therefore be incorporated.

Termination at the receiving end is usually by means of a transformer which prevents any d.c. being fed to the line. A 680 Ω resistor should be connected across the secondary which is fed to the slave amplifier input.

Ensure that the line is tested and operating well in advance of the meeting as it is not unknown for BT engineers to forget to make the link at the exchange. Chasing faults of this nature with limited time to go before the start of the event can be very exasperating.

One problem that does not exist with overflows is feedback as there is no microphone present, so a single pair of columns fed at high power usually gives good coverage. If on-the-spot announcements are required, then a mixer-amplifier will be required with a microphone.

Reverberant acoustics

These can be a real headache; feedback starts at an early level and intelligibility is poor. They are encountered in churches and cathedrals, swimming baths, arcades, and exhibition halls with high domed roofs.

The general treatment is to use a large number of loudspeakers run at low volume. This minimizes strong reflections, and all parts of the audience are fairly near a loudspeaker, thereby receiving a high proportion of direct sound from it.

In addition, directional loudspeakers should be used to aim the sound into the audience and away from the ceiling and walls. This is good practice for normal systems but is more so in highly reverberant situations. It calls for columns which should be elevated and angled sharply downward. A larger number than usual is thus required as the steep angles restrict the range, but this is desirable anyway for the reasons already noted.

If it is possible to acoustically treat some of the surfaces with absorbent materials, this helps. Curtains wherever appropriate are always a useful absorbent. As the loudspeakers are directing the sound downward, floor reflections can be significant but can be reduced by carpeting or matting. This is especially so if the floor is stone or tile. If the whole area cannot be covered, carpet runners in the aisles make a big difference, as these areas are exposed whereas the rest is partly covered by the audience.

Finally, tailoring of the frequency response with an equalizer may reduce some resonances and subjectively improve the intelligibility.

17 Test equipment

The test equipment needed by a public address installer depends greatly on the degree of his involvement, whether fitting up an occasional system for charity or local events, more regular installation work, or as a fully professional public address engineer. In our discussion of the various items we will assess their necessity for each of these classes.

Test meter

The multi-range test meter is an indispensable item for all installers whatever the amount of their involvement. Basic ranges measure a.c and d.c. voltage, d.c. current, and resistance. There is a tremendous choice of instruments, from the cheap pocket unit to the expensive professional one. The differences are in the number and extent of the ranges, sensitivity, accuracy, size of read-out and robustness. There is also a choice between digital and analogue.

Taking the last point first, digital meters are more accurate and robust than analogue. They also impose a lower load on the circuit under test. For example, a 20 kΩ/V analogue movement places a load of 2 MΩ on the 100 V range and only 2000 kΩ on its 10 V range. The current drawn produces a voltage drop in a high-impedance circuit, which therefore gives a false reading. If the circuit impedance is the same as that of the meter, the voltage reading is halved.

The standard input resistance of a digital meter is 10 MΩ on all ranges. So the current drawn is negligible and the voltage reading, even with a high-impedance circuit is not far short of the actual value.

Nevertheless, the impedance of most transistor amplifier and other circuits encountered in public address work is not very high, so the inaccuracies of an analogue meter are not usually serious. Cheap analogue meters having a resistance below 10 kΩ/V are correspondingly less accurate, but there are some models having a resistance as low as 2 kΩ/V which are prone to large inaccuracies.

Apart from inaccuracies due to internal resistance, digital displays are more accurate than mechanical ones. For $3\frac{1}{2}$ digit displays accuracies of around 1% are usual for a.c. volts and 0.5% for d.c. With $4\frac{1}{2}$ digit displays 0.5% for a.c. and 0.05% for d.c. are common. In the case of analogue meters 4% for a.c. and 3% for d.c. are general.

However, for public address work extreme accuracy is not a high priority. Many measurements are comparative, so as long as they are made with the same meter, a few per cent error is of little consequence. Voltages given in service manuals usually have a wide tolerance due to component tolerances, so again, strict accuracy is not essential.

A feature of some digital meters is *auto-ranging*. With these the voltage range does not have to be selected by switching, it adjusts itself to the voltage being measured. This can be useful when fault-finding in amplifiers and consecutive measurements are made at various points which may have widely differing voltages. It is also useful when measuring unknown voltages as it saves switching down through the ranges from the highest, as is the normal practice.

Another feature is the *display hold* facility which holds the reading after the instrument has been disconnected. With conventional meters the prod must be held on the contact point while the scale is read. With attention thus diverted from the prod it can slip off, and so easily short against an adjacent contact, with possible catastrophic results. Display hold allows full attention to be paid to the prod, then the measurement read after the prod is safely removed.

Many digital meters have other ranges besides the normal voltage, current, and resistance. Capacitance is one, though this is usually rather limited at both ends. It does not go low enough into the picofarad range to be useful for cable-break testing (see Chapter 18), nor high enough to measure the high-value electrolytics commonly found with modern amplifiers. A separate capacitance meter is thus to be preferred.

Some have a frequency counter which extends over the whole audio spectrum and beyond. This can be useful for setting up an anti-feedback equalizer by determining feedback frequency, but would not be needed for adjusting variable notch filters as these quickly tune out peaks without the need to identify them. Resonant peaks in equipment or auditoria could be determined by using a frequency counter, and this could assist in the design of suitable filters. A frequency counter is thus a useful facility for the professional.

Other multi-meters incorporate a transistor tester giving h_{FE} measurement. Useful, but not essential, a separate tester will give more comprehensive checks.

While analogue meters normally lack these features (although some do have capacitance and h_{FE} ranges) and are less accurate, they are still generally to be preferred for public address work. Unlike most digital instruments they can indicate a varying signal. They can thus be used as an output meter, to measure signals on loudspeaker feeders, output from mixers, programme sources, and other signals. The digital meter just produces a rapidly changing jumble of figures which is quite incomprehensible.

A few digital meters include an analogue bargraph display which does enable the instrument to be used for varying signal measurement. The scale tends to be small and not easy to read, so it very much a comprise.

Some analogue meters have a *centre zero* feature by which the pointer can be zeroed at the centre of the scale. This enables measurements to be made without regard to polarity and so eliminates the need for changing round the leads for different tests. It is also useful for measurements in bridge circuits and push–pull stages where readings can vary from positive to negative.

Another feature with many analogue meters that is not found with the digital variety is a decibel scale. This enables comparative measurements to be read off directly in dB when signals are measured on the a.c. ranges. This can be very useful for many public-address applications.

The upper voltage ranges, both a.c. and d.c., are more than adequate with even the cheapest meter, being above 500 V. It is the lower end that needs more scrutiny. A 10 V full-scale deflection (FSD) for a.c. is common, but a lower range, at 2.0 or 2.5 V FSD is desirable for measuring low signals; better still a millivolt range, although these are rare.

Current ranges usually go very low, 50 μA FSD for a 20 kΩ/V analogue movement, which is the current that passes if 1 V is applied. Digital meters generally do not go as low as this. However, in this case it is the higher range that is important. The output stages of high-power amplifiers pass high currents and some means of measuring them should be available. A test meter should therefore have a high current range of not less than 10 A.

An a.c. current range is usually found only on the more expensive instruments, but it is highly desirable for checking mains current to equipment, and loudspeaker feeder currents among other things.

Resistance ranges differ considerably between models. The upper range should extend to 10 MΩ, preferably 20 MΩ. Not that resistances of this order are often encountered in public address work, but analogue resistance scales are cramped at the high end, so a high range is needed to enable accurate readings of lower values to be made nearer the middle of the scale. To give a readability indication, some makers specify a midscale value of each resistance range.

The lowest range should also go well down, as low reading measurements are often made. A 0–1 kΩ range or lower is desirable; with ranges much above this, resistances of a few ohms will be indistinguishable from dead shorts.

It is with resistance ranges that digital meters show their worth. The lowest range often is 200 Ω, and accuracy is usually better than 1% compared with the average 3% of the analogue meter. This enables resistors for critical circuits to be hand selected for value.

Summing up then, while some form of multi-range test-meter is essential for anyone engaged in public address work, the type and cost depends on the nature of the work, whether repair and servicing of amplifiers is undertaken, or only system installation. It is a good practice to choose a slightly better model than you think you will need, as unexpected uses are often found for facilities at first considered unnecessary. The professional or frequent user may find both an analogue and a digital meter useful for their respective features.

Audio signal generator

This instrument produces sine waves within the audio range. Frequency is selected by range and continuously variable controls within each range. Amplitude is also variable, usually with both coarse and fine controls. Some models give an output to 1 V, others up to 10 V. The former serves for most applications, but the latter are useful when injecting signals into high-level circuits such as driver and output stages.

Ideally, the signal amplitude should be maintained over the whole range, so that frequency-response tests can be made without the need for measuring and resetting the generator level at each frequency. However most moderately priced instruments fall off in output at the extremes of their range. Some have a built-in meter so that the output can be adjusted to give a required output level at all frequencies.

Many generators also produce square waves. As a square wave consists of a sine fundamental with an infinite number of odd harmonics, a display on an oscilloscope immediately reveals the frequency response of an amplifier, without many repeated measurements at different frequencies. The display will show up any parasitic oscillation in an amplifier. The square wave facility is therefore a very useful one.

The generator can also be used as a signal source for testing complete installations or making phasing checks. Live speech, though, is the best ultimate test as intelligibility and feedback are thereby checked.

The generator should have a reasonably low distortion, but not to an extreme. Low-distortion models are essential for making distortion measurements on hi-fi amplifiers, but this is not a consideration with public address equipment. Such models are expensive and are not necessary for this type of work. Often the low-distortion models are termed *generators*, while the ordinary service models are called oscillators, although this distinction is by no means universal.

For the public address engineer an audio generator is essential, especially if the repair of equipment is undertaken. The installer will find it of less value although occasionally useful. Inexpensive battery-operated generators are available which may serve the purpose in such a case. Many of these are surprisingly good.

Sound level meter

This is an instrument that contains an omnidirectional microphone, amplifier, and a meter. A range switch permits readings between wide limits, typically 40–110 dBA. Another switch selects weighting curves; on simpler models these are between a flat response and the *A* curve. Those having no switch have an *A* curve response.

Another control enables a fast or slow meter response, the latter being chosen to give an average reading when sound variations are too rapid to follow. Some models have a hold facility so that the highest reading is held.

The presence of an omnidirectional pressure microphone in the sound field itself modifies the field and can cause erroneous readings at high frequencies. Some models are velocity microphones that do not disturb the field and so give more accurate readings, but these, being directional, must be pointed at the sound source. They are therefore inaccurate in a diffuse field as they do not respond to reflected sounds coming from other directions.

The smaller the diaphragm, the greater the omnidirectional characteristic of the microphone. Sound-level meters therefore use microphones that are as small as possible without sacrificing too much sensitivity, to make them non-directional.

With diffuse fields, sound can be blocked by the operator and even the case of the instrument. Furthermore, reflections can be produced by the same causes. At frequencies around 400 Hz, reflections from the body of an operator can cause errors of to 6 dB. To reduce these effects, the microphone is mounted on an extension rod in front of the instrument and the instrument front is made conical in shape.

These details are features of the more expensive professional instruments designed for very accurate sound measurement for noise investigation and similar applications. For public address work the requirements are less stringent, and one of the more inexpensive models can be adequate. These do not have conical fronts or small microphones on extensions and so are not accurate around 10 kHz and above, but this frequency region is not important for public address speech.

One thing to watch out for is the number of ranges. One model has its total compass divided into only two ranges, and as a result has a compressed scale on which it is not easy to distinguish small differences. Another has six ranges with a correspondingly more open scale.

There are many uses for the sound-level meter. Checking for an even level over an auditorium is the most obvious. Approximate readings can be taken on speech when a session is in progress, measuring peaks and ensuring the level is adequate everywhere. A more accurate method when making initial tests is to use pink noise. This is random noise in which an equal amount of energy is contained in each octave.

With areas of high ambient noise, reproduced speech should be 10 dB higher in order to be intelligible. Using the meter, noise level can be first measured to determine the sound level required.

Checking the phasing of units in a column or LISCA array reproducing pink noise or a fixed tone, can be carried out by running the meter along it at a distance of a few feet. An appreciable dip in level at any point indicates either that a driver is not working or it is connected the wrong way round. However, it should be noted that if a fixed tone is used the dip could be the result of interference from some reflection. Pink noise is preferred. In similar manner an unexplained dip in level in an auditorium could be due to one of the loudspeakers serving that area being out of phase.

The effectiveness of various materials for sound insulation at different frequencies can also be determined by tests with a sound-level meter. It is then an altogether worthwhile instrument for the public address installer and engineer alike.

Pink noise source

Noise is used for audio tests in place of single tones because being a random collection of frequencies over a wide band, any interference effects or resonances at certain frequencies are greatly diluted. Thus accurate sound level measurements can be made in halls and other enclosed areas where fixed tones would lead to errors.

White noise is a random band of frequencies In which half the energy is concentrated in the highest octave (8–16 kHz), and a quarter in the next (4–8 kHz). Three-quarters of the total energy is thus above the band of speech frequencies. This is why white noise sounds 'hissy' when it is heard as inter-station noise on f.m. radio—it mostly consists of high frequencies. White noise can therefore be misleading if used for measuring sound levels in a public address system; it does not accurately convey the effect that speech has through a loudspeaker system.

As all moving-coil loudspeakers have a strong beaming effect at high frequencies (typically 25° off-axis for -6 dB above 3 kHz), any readings taken off-axis are low when white noise is used.

Furthermore most absorbent materials have a large absorption coefficient at high frequencies. Measurements using white noise are therefore affected by the proximity of absorbent surfaces such as chairs with padded seats, and carpeting.

For these reasons white noise can only be used for rough checks, but never for quantitative or comparative measurements. For this, pink noise is used, in which the energy is distributed equally in each octave. It has thus a much lower high-frequency content and is not so affected by absorbents. Having a greater proportion of speech frequencies it relates more closely to the effect of speech through the system.

An example of the difference in off-axis readings is that a typical 6in (15 cm) drive unit has a 70° −6 dB point measured with pink noise, but only 50° with white.

These points are stressed because some installers are known to use white noise for testing due to the ease of obtaining it. The source is usually a detuned f.m. radio, or a tape recording made from it.

A pink noise generator, which produces filtered white noise is a useful item in the professional engineer's workshop. A less expensive means of obtaining pink noise is from one of the various test records that contain it. A convenient method is to record from one of these on to an endless tape cassette. This can then be played on a portable recorder plugged into the system or on the system's own tape deck if there is one.

The endless tape cassette consists of a short length of tape in a standard cassette that is formed into an endless loop. The advantage is that it can be played for an unlimited period, so pink noise is available for as long as the tests take. If recorded on an ordinary cassette, the recording invariably comes to an end just before the tests are finished, whereupon a journey back to the control point has to be made and the tape rewound. With endless tapes though care must be taken to insert them in the recorder the right way—they only play one side.

Impedance tester

This is a device that measures the value of an impedance by passing through it an internally generated a.c. current at a standard frequency. The scale is calibrated directly in ohms. Ranges usually extend from 10 Ω FSD up to several thousand ohms.

As impedance varies with frequency, any reading holds good for only one frequency. There is therefore normally a choice of internally generated test frequencies, as well as provision for using an external one. Impedance of audio devices such as loudspeakers and microphones is generally measured at 400 or 1000 Hz.

The tester can be used for checking the impedance of unknown loudspeakers and those with line transformers having unmarked tappings. The power rating of these with a 100 V system can then be determined from the measured impedance by 10000/Z, where Z is the impedance.

Total loading on a loudspeaker feeder can be calculated in the same way after measuring its impedance. This can be very helpful, especially in a large temporary installation when others have been involved in connecting the loudspeakers. Transformers may not have been correctly tapped, but comparing the total calculated load of the feeder with what it should be soon reveals any discrepancy.

On completion and testing of any large installation, the impedance of each feeder can be measured and noted in the service data. This information can be useful when tracing faults in the future. Especially with a large temporary system in which loudspeaker feeders are vulnerable to damage, a check and comparison of feeder impedances well before the start of each session may forestall trouble.

The impedance tester is thus a useful device, especially to anyone installing large systems and dealing with unknown equipment, but perhaps less so to the smaller contractor.

Capacitance meter

At one time few workshops were without a bridge for measuring capacitance and resistance. The device consists of four impedances connected as a square. A power source is connected across two diagonal corners, and a meter or some other indicator across the other two. When all impedances are equal there is no potential difference across the second pair of corners and so the meter reads zero and the bridge is said to be balanced.

If one impedance is unknown, the bridge can be balanced by changing one of its adjacent impedances by the same amount. If this impedance is made variable and calibrated, then the value of the unknown one can be determined by adjusting the other for a null reading on the meter and reading off the calibration.

Ratios between the arms of the bridge can be varied by changing the values of the other impedances, and this enables the basic range to be extended.

Most bridges measure resistance and capacitance, but accurate digital multimeters have removed the need for the former, and direct reading capacitance meters have now come into vogue to render the latter function obsolete too. In both cases, the meters have a greater range than the bridge, and are more convenient to use.

A capacitance meter can read from a few picofarads to several thousand microfarads and so encompass most of the values likely to be encountered. As noted before, those built-in to digital multi-meters tend to be rather restricted in range.

Next to semiconductors, capacitors are the components most likely to give trouble, being far more vulnerable than resistors. So some means of testing them is essential to anyone engaged in servicing equipment. Varying capacitance which heralds an incipient breakdown can be observed on the meter, which can detect leaks as well. Apart from components, a capacitance meter is very useful in detecting and localizing cable breaks.

The bridge has therefore fallen from favour for *RC* measurements, but those bridges that also measure inductance could be of value as there appears to be no other instrument that reads inductance directly. Although inductance measurements are not often needed in public-address work, the facility could be of assistance in checking inductors in equalizers that use them, or for r.f. interference and other filters. It would not be worth while for the occasional installer, but is a possibility for the engineer.

Insulation tester

Insulation testers, also called meggers, are resistance meters that measure high resistances of many megohms. They do this by using a high voltage source which is electronically generated within the instrument. The old megger used a hand-cranked generator. Voltages of 250, 500 or in some cases 1000 V are employed.

Voltages of this order are used to break down faulty insulation. A measurement using a multimeter may not show up a leak even though the meter can read into the megohm range. This is because the meter applies only a low voltage to the circuit. Poor insulation will often hold at low voltages, but break down at high potentials. The insulation tester tests the ability of the circuit to withstand high voltages as well as indicate the leakage resistance.

Care must be taken, especially with the 1000 V models, that the item under test is designed to stand such a voltage. Most capacitors are rated below 1000 V, as are most cables. Electrical wiring cables are generally rated at 500 V between conductors and 300 V between conductors and the earth wire. Testing at 1000 V could therefore produce leakage readings and in some cases insulation breakdown. A tester voltage of 500 V is the best rating in most cases.

An insulation tester could be useful for large installations, especially outdoor ones, but there is little call for it for most public address work.

Transistor tester

For all engaged in amplifier and equipment repair, a transistor tester is a must. These enable the h_{FE} to be measured and leakages checked for most bipolar transistors, while some models include admittance tests for FETs. Of particular value is the ability to match push-pull transistors in output and driver stages. A mismatch ratio of 2:1 can produce second–harmonic distortion of 16.5%, and h_{FE} difference ratios of several times can often be found between two samples of the same type of transistor.

While distortion so produced is reduced by negative feedback, it is desirable to match replacement output transistors as closely as possible, although a close match is rarely attainable from a few samples. Selection of the closest pair though is made possible by measurement with a tester.

A.c millivoltmeter

A.c. voltages in the millivolt range are not usually available on testmeters except amplified meters with FET inputs. The main problem is the forward voltage needed to turn the rectifiers on; for silicon, this is 0.6 V and germanium 0.1 V. Other than moving-iron meters which read a.c. direct, a linear voltage amplifier is required to precede the rectifier.,

Millivoltmeters are necessary for small-signal measurements with preamplifier stages and signal sources. They are often built-in to audio analysers and distortion meters and can sometimes be accessed for external measurements. Again, this is useful for the professional engineer, but a system installer may find little use for it.

Oscilloscope

A most versatile and useful instrument. A wide range of models now exists but many have features that are not required for audio work. For example, *Y* amplifier frequency response of 50 MHz and more is available, and while it may be essential for television servicing it is not necessary for audio. Similarly the *X* sweep frequency can extend into the megahertz region, but up to 100 kHz is sufficient to display audio waveforms.

Although not essential, a double beam facility is a help when two waveforms are to be compared or monitored simultaneously. Some models have a freeze display whereby part of a varying waveform and be stored and held. This too has its uses especially for design and development work, but has limited application for public address work.

It follows that quite a basic and inexpensive model serves well for most audio applications and so can be part of the kit of most workers in the field, as well as professional engineers. Many faults can be identified by its use. Overloading and clipping can be observed before it becomes audible, various types of distortion can be indentified as well as incipient faults such as instability.

A scope connected to the output of a mixer makes an excellent monitor. Many of the above faults can then be instantly diagnosed, while levels can be seen at a glance and maintained within the required limits.

There are two points to watch for. One is the vertical sensitivity. Cheaper instruments may be lacking in this respect, but good sensitivity is required to observe low signal levels and any distortions of their waveform. A sensitivity of 1 mV per graticule division is desirable, but 2 mV is acceptable. Anything much more than this will limit the instrument's usefulness at low signal levels.

The other point is screen size. Some instruments have tiny screens of less than 2 in (50 mm). This is insufficient to observe waveform subtleties, the minimum screen dimension should be 3 in (76 mm).

18 Faults and their cure

One of the worst headaches for any public address engineer is for a major breakdown to occur during a big event with a large audience. Should this unhappy situation arise never panic (at least not to start with!), always adopt a cool systematic fault-finding procedure. Never be pressured into a hit-or-miss approach; observe symptoms, draw conclusions and eliminate possibilities. This is easier said than done, but it is the only way of ensuring that the fault is eventually traced and rectified.

The various monitors described in Chapter 12 are invaluable as they can lead to a quick isolation of the defective item or circuit. Having done that, the next step is to decide quickly what emergency measures are needed to restore as full a service as possible in the shortest time. Attempting to repair the faulty item there and then as a means of getting things going is rarely practical. Most repairs take time to do properly, and with a scheduled programme and a waiting audience, that sort of time is just not available.

The solution is to have standby equipment on hand so that a quick changeover can be made, even if it is at the cost of reduced facilities. Of course a repair can be tried when the back-up is installed and running. If successful, the repaired item can itself then serve as a standby, or can be put back in service during a suitable interval.

Some common faults are listed in the following sections.

Partial loss of sound

Either the entire system has gone dead or only a section. In the latter case it is obvious that the input circuits are not responsible, so microphones, mixer, and their leads are immediately eliminated.

The fault must lie either with the amplifier supplying the affected area or its loudspeaker feeder. The output meter or monitor will indicate whether the amplifier is working, but before jumping to the wrong conclusion, the feeder should be disconnected as a short-circuit would remove the signal.

The feeder is the most likely cause of the trouble especially with a temporary installation. If it is open-circuit, some of the loudspeakers may be working, thereby indicating the point at which the break has occurred. If all are dead, the break must be between the control point and the first loudspeaker, or there is a short-circuit. A resistance check with the meter will soon reveal which.

There may be quite a long run between the amplifier and the first loudspeaker, and discovering where a short-circuit is may take a careful visual examination of the feeder and its connections, which could be a lengthy process. The location can be considerably narrowed down by noting the exact resistance obtained across the feeder. If twin 16/.02 cable is used, this has a resistance of 6 Ω per 100 yards for both conductors. If then the measured reading is 2 Ω, the short is about 33 yds along the feeder. If working in metres the figure is almost the same.

An open circuit is not so easy to find but a similar method can be employed by the use of a capacitance meter. All cables possess the property of capacitance due to the proximity of the conductors, and this is proportional to the length. Thus the length to the break can be calculated from the measured capacitance.

No exact figure for the capacitance of cable can be given because it varies according to the thickness of insulation, and in the case of twisted cables, which are recommended for loudspeaker feeders, whether the twist is tight or loose. It is approximately about 30 pF per yard. A good idea is to measure the capacitance of a short length, say 10 yards or metres of the cable used, in advance, and note the value for future use.

Should the amplifier prove at fault, perhaps the feeder can be temporarily plugged into another that has sufficient spare power capacity, while a standby is prepared or the fault investigated. It is a wise precaution with a large installation having several amplifiers, to have one lightly loaded so that it can take on an extra load in an emergency.

If this then takes up all the spare power capacity, it is prudent also to investigate the faulty amplifier to see if it can be repaired, because there is then no standby should a further breakdown occur. It is rather like having a puncture in the spare wheel!

Total sound loss

This is the one to cause panic. However, the fault is almost certainly nearby and usually easily corrected. All the amplifiers and feeders are absolved from blame as faults with those affect only one section. There is one exception, and that is supersonic instability, which could generate an inaudible high-level swamping signal which could be affecting all amplifiers. In this case all the output meter needles will be hard over. If there is no such indication, that is not the cause and the fault is elsewhere. We will deal with instability later.

The possible causes are: the microphone, microphone leads, the mixer, mixer lead to the amplifiers, power supply. The latter will be obvious from the power indicator lamps on the equipment.

Leads are always the most likely causes, so the chief ones here are the microphone lead and the mixer lead. The mixer output meter or monitor will reveal if the signal is getting that far. The problem here is that if the system goes dead, the speaker, realizing he is not being heard, usually stops talking and looks inquiringly at the public address control for confirmation to continue. He thereby deprives the operator of a 'test' signal. Here is where an oscilloscope connected as a monitor to the mixer proves its worth. It shows ambient noise that is always present, at a glance, whereas the noise may be too low to register positively on an output meter. Headphone monitoring can achieve the same result.

For important occasions it is wise to have a spare mixer lead laid out alongside the operating one. If the monitors show that the signal is reaching the output of the mixer, the lead can be quickly changed and the sound should then be restored.

If the mixer monitors are dead, most likely the fault is with the microphone lead. Another lead with microphone should be ready plugged in as a standby, or maybe there are others in occasional use. Fading up one of these should produce ambient noise on the monitor. If not the mixer itself is at fault. This is unlikely but it can happen. A standby, even if only basic, will get the programme going again even if some facilities are no longer available.

Microphone lead faults

The most common faults with microphone leads are breaks, and these nearly always occur at the microphone end within a couple of inches of the connector. The obvious treatment is cutting off a few inches and reconnecting. Seldom is the break inside the connector. Before doing that though it is as well to confirm that the fault is there; though less likely, the break could be at the other end.

Here the capacitance meter will again prove a big help. There is no need to know the capacitance of the cable. When measured it will either be a significant amount or zero. If it is zero, that is the end where the break is, if not, it is at the other end. If in any doubt measure from the other end. If there should be a sizeable reading from both ends, the break is somewhere along the cable, and the ratio between the two readings will give the approximate position. This is rare, but can happen.

The cause of microphones cable breaks is repeated flexing during use at the point of entry into the microphone connector. It is reduced by means of a strain reliever which is either a long rubber grommet of reducing thickness along its length, or a spiral spring. While these reduce the incidence of breaks they do not completely prevent them, they usually just shift the break to the end of the strain reliever.

The best prevention is using a suitable cable. The ordinary single-conductor screened variety is very prone to breakage and should not be used for microphones even with unbalanced systems. Round twin screened with fibre or string padding is the best. This does not develop sharp bends or kinks, and it descends from the microphone in a gentle curve, so avoiding continued flexing at one point. A dramatic drop in the incidence of lead troubles is always experienced when changing to this type of cable. A further advantage for unbalanced systems is that the quasi-balanced arrangement described in Chapter 6 can be adopted.

Noise

Noise can take many forms, but it is usually heard as a continuous hiss or as tearing and crackling noises. It is often due to a fault in the mixer, as noise in power amplifiers is usually swamped by the high signal level from the mixer. If of a sufficiently high level, though, noise in an amplifier could intrude over the signal.

Headphone monitoring of the mixer is best for revealing whether it is the culprit, as the operator can thereby distinguish between circuit noise and ambient noise picked up by the microphone.

If the mixer is at fault, it is probably only one input that is the cause, so the microphone plug can be quickly changed to another input during a suitable pause in the proceedings. If there is no difference, a later stage of the mixer must be to blame. If the noise is not too intrusive it may be best to wait for an interval before changing to the standby mixer, rather than interrupting the programme. Turning down the treble control in the meantime will help in the case of hiss, as it is mainly composed of high frequencies.

Tearing and crackling

Tearing and crackling sounds are usually due to a component breaking down under voltage stress. It is most likely to happen in a power amplifier as heat is often a contributory factor if not the cause, so checking the output monitor should reveal which. If all are affected then the fault must be in the mixer.

Faulty volume or fader controls are frequently the cause of crackling noises. Movement of the control soon confirms whether this is the case. The trouble is sometimes due to dirt on the carbon track, but is more likely to be oxidization of the sliding metal contact. It can also be a worn track, the only cure for which is replacement of the control.

Dirt or oxidation can be cleaned off with a switch cleaner applied through apertures in the back of the control, or if there are none, by prising off the cover after bending back the securing tabs. This loosens the dirt or oxidation, whereupon rapidly rotating the control of sliding the fader up and down will in most cases clear it.

Many of the commercial switch cleaners are a mixture of a cleaner and a lubricant and they do not always produce the desired result in stubborn cases. If the noise persists, try an application of carbon tetrachloride, followed by a little light oil. Avoid inhaling the vapour, and keep the fluid away from the plastic parts as it dissolves some plastics.

If there is no cleaner or carbon tetrachloride to hand, the oil by itself often does the trick. If even this is not available, or the noise appears at an inopportune time, try rapid rotation or sliding, pulling the knob towards you to exert more pressure on the sliding parts. This sometimes improves matters as a temporary measure.

Radio breakthrough

The proliferation of public service radio communications, pirate radio stations, and CB transmitters which often exceed the legal output power, has increased the possibility of interference with public address systems. The result is that disturbing background music or bursts of speech occasionally breakthrough.

The cure can in some cases be straightforward, but in others, especially if there is a powerful transmitter nearby, it can be difficult to achieve. To pick up radio transmissions an aerial is required, which in its simplest form is a length of wire. The aerial is most effective when its length is a quarter of the wavelength of the radio wave being received, or an exact multiple of it.

In addition, a tuning circuit is needed to eliminate all the unwanted stations picked up by the aerial, leaving the wanted one. Then, the radio signal must be demodulated. An a.m. radio signal has audio modulations on both its positive and negative halves, which, being in opposite phase, cancel. The demodulator eliminates one half of the radio wave by rectifying it, usually with an r.f. diode. This leaves the audio modulation to be amplified by the audio stages.

In most public address systems all the above elements are present except the tuning circuit, which means that often two or more stations are received simultaneously. The microphone cables serve as an aerial; the input circuit of the first transistor stage (base–emitter junction) passes current in only the direction and thereby demodulates the radio signal. The carrier is filtered out by the treble roll-off of the following circuits, leaving the audio modulation, which is amplified by subsequent stages.

The simplest remedy is an input bypass capacitor. Long microphone leads pick up radio signals, so a non-inductive (ceramic) capacitor of 0.015 μF, connected across each microphone socket, can bypass much interference picked up on the cable between the microphone and the socket. There is no need for high-voltage capacitors—25 V is quite adequate. All connections should be soldered.

If there is a long run from sockets on the platform to the mixer, another capacitor should be wired across the mixer input sockets. It may be possible to fit the capacitors inside the plugs; low-voltage capacitors of this value are quite small. When two capacitors are thus used on the same run they should each be of 0.01 μF capacitance.

Interference may be experienced on only one cable or one input channel. The cable could be of a critical length, being a multiple of the quarter wavelength of the interfering station. Try shortening it by a foot or so. It may still need bypass capacitors, but its efficiency as an aerial at the interfering wavelength may be reduced.

Very long microphone cables are best shielded. This can be done by running in electrical steel conduit, or copper central-heating tubing. The latter may be more effective and easier to work with. In either case, the tubing should be well earthed, preferably by a metal spike driven into the ground. Mains earths are often poor for interference shielding. Do not use long multiple-core screened cables to take all channels, they are more prone to radio interference than individual cables. Do not use television coaxial aerial cables as these have poor screening at audio frequencies.

In obstinate cases when the above measures fail to eliminate the problem a

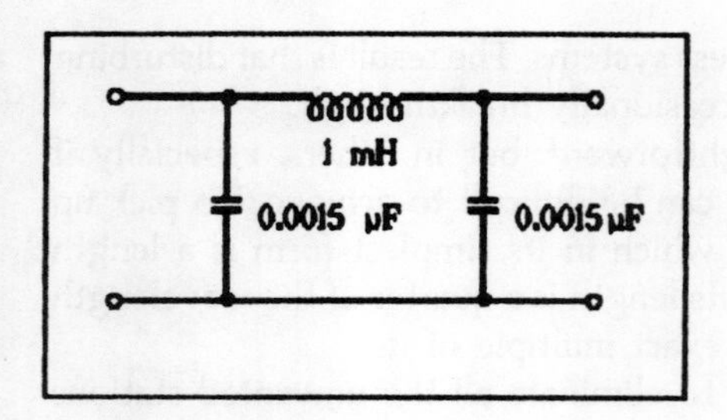

Figure 94 Pi microphone input filter for radio interference.

pi (π) filter may be required. This consists of a capacitor across the input as before, plus an r.f. coil in series with the live side of the cable, and another capacitor across the cable as shown in Figure 94. The capacitor value is 0.0015 μF each, and the coil, 1 mH (Maplin WH47B). All joints should be soldered.

MOSFETs do not pass input current in any direction and so do not demodulate radio signals. A mixer using these for the input stage should therefore not suffer from radio interference. However, they are not generally used for p.a. input stages, although some models may have them. MOSFET output stages make no difference to radio interference.

Microphone sockets and plugs should be kept clean and contacts smeared with a little light grease. Tarnish can behave as a diode, and so a dirty contact could demodulate radio signals prior to the bypass capacitor. Systems should be well earthed, but to only one point. So if separate mixer, tape deck and amplifiers are used, only one, preferably the mixer, should be earthed. Very rarely, radio could be picked up on the speaker wiring, and jump to the microphone inputs at the amplifier. Test for this by disconnecting all speakers except one and give a long test run.

Hum

There are many possible causes of hum, so a process of elimination must be employed. If it appears only one amplifier out of several, the amplifier itself is almost certainly the cause and should be taken out of service at a convenient interval for repair. Internal causes of amplifier hum will be dealt with later.

If hum is present on the whole system it could be due to an earth loop, an open-circuit earth connection, disconnected braiding of a screened lead, or careless positioning of equipment units so that the input circuits of one are adjacent to the mains transformer of another. Microphone transformers are especially prone to hum pickup unless screened in mu-metal, but these are not often used with modern semiconductor input circuits.

An earth loop is formed when there is more than one earth connection to the system. If, for example, both the mixer and the power amplifier are earthed, there is a continuous path from the one earth to the unit case, along the braiding of the screened lead between them to the other unit case, and from there back

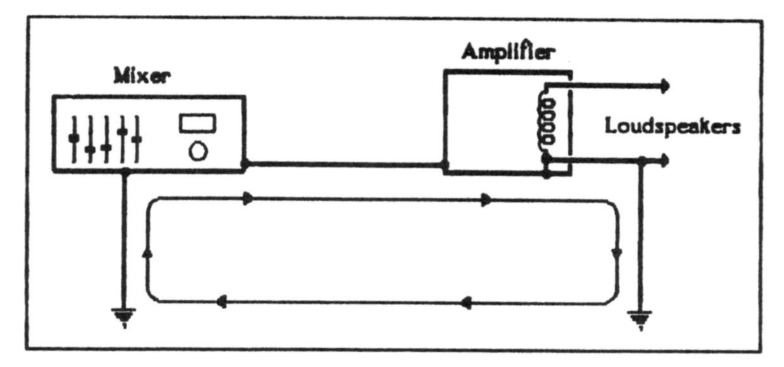

Figure 95 An earth loop caused by an earth leak from one of the loudspeaker wires. Induced hum currents can then circulate through the mixer lead braiding.

to earth. Any hum field through which part of the loop passes will induce hum currents that will circulate around the loop. As these pass along the braiding which is effectively in series with the amplifier input circuit, hum is thereby introduced. This is the reason why the advice has been previously given to earth only at one point (see Appendix).

However, a second earth connection could be made accidentally. Where one side of the loudspeaker system is connected to the amplifier case, a short-circuit from a loudspeaker feeder to earth will effectively earth the amplifier and so produce the loop described. Figure 95 shows the path.

Earth loops can be baffling because disconnecting anything in the loop path breaks the loop and stops the hum. So, in the above case, while disconnecting the microphones would have no effect, unplugging the mixer would, thereby suggesting that the mixer was at fault. Trouble with the loudspeaker feeder would probably be the last thing suspected.

Hum caused by earth loops is usually pure-toned and not very loud, although it can be intrusive. As earth loops are probably the most common cause of hum, especially with a temporary installation, it is as well to check for a loop if hum of this nature appears.

The quickest way is to disconnect the units from each other, but not the loudspeaker feeders, then take a resistance check, preferably with an insulation tester, from the case of each unit to a known good earth. If an earth is found on an amplifier, disconnect each loudspeaker feeder in turn to find which is responsible.

Another unusual, but possible, cause of hum is an open-circuit or poor earth connection. One method of testing is to pass a current through the earth connection and monitor any voltage drop. The easiest way of doing this is to connect a 100 W lamp from the live side of the mains to the suspected earth connection and measure the voltage across if (Figure 96). The lamp should light at full brilliance and the meter read the full mains voltage. Any voltage drop

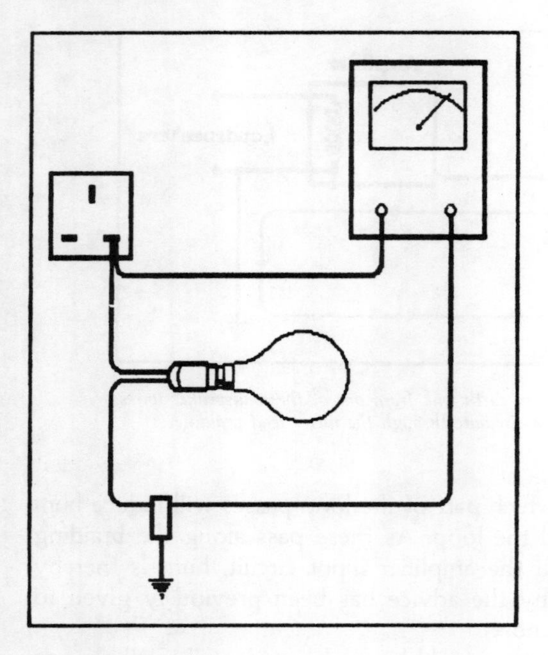

Figure 96 Checking an earth connection with 100 W lamp and meter.

indicates series resistance. Actually, the earth connection should be capable of passing a much greater current than the less than $\frac{1}{2}$ A of the 100 W lamp, but it does provide a quick and convenient indication when checking for hum. This test is not possible when an earth trip is in circuit.

Open-circuit screening on a microphone lead will produce a loud hum with a balanced circuit and an even louder one with loss of signal with an unbalanced one. As the normal signal is small compared to the induced hum voltage, the hum usually causes overloading and clipping of its peaks. The result is more like a hard buzz than a hum.

Early feedback

This is due to poor layout of loudspeakers, or the use of non-directional microphones, or those with response peaks. Reflective untreated walls also produce feedback. The subject is fully dealt with in Chapter 11.

Distortion

As with previous faults, the first step is to check whether the trouble is confined to one amplifier or is present over the whole system. The most likely source of distortion is the power amplifier, so it would be expected that just one amplifier would be affected. If as suggested in Chapter 12 a loudspeaker amplifier monitor is switched around the amplifiers from time to time, distortion in any one will be detected at an early stage.

While the trouble could be an internal amplifier fault, possible external causes must not be overlooked. One likely one is a partial short-circuit across a loudspeaker feeder. This causes a serious mismatch, distortion and low volume. Unplugging the feeder with the monitor still on to see if the distortion clears reveals whether this is the cause of the trouble.

Mixer faults usually cause noise or loss of signal rather than distortion. If there is distortion on all the amplifiers, the mixer is a possible cause. Another possibility is supersonic instability. This may affect the whole system and be inaudible, but it can drive the amplifiers into overload, thereby producing distortion. It registers on the output meters as a steady reading even when there is no input signal, often with the meter needles hard over.

Overloading due to too high a signal being applied to an input, either mixer or amplifier, is a common source of distortion. It is shown up as clipping of the waveform on an oscilloscope.

Microphones can produce distortion, especially ribbons which are rather fragile. Usually the trouble is as a result of physical shock, which displaces the ribbon and causes it to rub on the magnet pole-pieces.

Instability

Instability occurs when a portion of the output signal finds its way back into earlier circuits. It is re-amplified to the output, fed back again to the earlier stage, and so on until in a fraction of a second a large spurious signal has built up and oscillation takes place. Theoretically it is similar to acoustic feedback except that sound is not involved in the feedback chain. Sound may be produced in the form of a squawk or whistle, but as noted above it could be supersonic, and it could drive the amplifiers into distortion or even paralysis by overloading. The steady reading on the output meters with no signal is the sure symptom apart from any audible oscillation.

There are three possible ways that the signal can find its way back to the earlier stages; *electrostatic, electromagnetic,* and *common-impedance* coupling.

Electrostatic coupling arises when input and output components or leads are in close proximity, thereby producing a capacitance between them. Instability due solely to capacitance is unusual at audio frequencies because of the low values of capacitance involved, but it can take place at low radio frequencies if the bandwidth of the amplifier is large enough. To eliminate this possibility, the amplifier upper frequency limit is often purposely curtailed to roll off just above the audio range.

A more likely cause is electromagnetic coupling, produced when the magnetic field from an output transformer or loudspeaker feeders interact with input cables

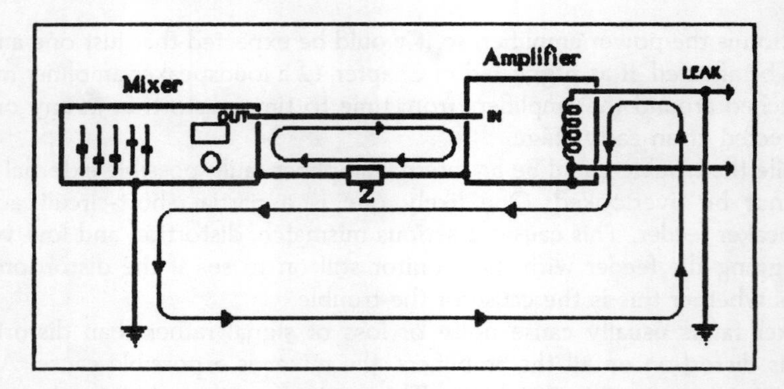

Figure 97 Common impedance coupling caused by a leak to earth from a live loudspeaker lead. Output currents flow along the mixer lead braid which is common to the input circuit. Instability results.

or components. This could happen when loudspeaker feeders and microphone cables are run for any distance in close proximity. Turning down the microphone faders in turn will identify which is responsible. Prevention is better than cure so it is prudent to keep microphone and loudspeaker feeders as far apart as reasonably possible.

The common-impedance coupling occurs when some impedance is common to both input signals and those at higher amplification, usually the output. The high-level signal produces a voltage drop over the impedance which is thereby communicated to the input circuit. If the signal is in phase with the input signal or nearly so, instability results.

The impedance does not have to be a resistor or inductor; a piece of wire or the braid of a screened cable possesses inductance and resistance, which though small could be sufficient to produce enough coupling at high gain to give rise to instability. Reference to Figure 97 shows one example of how such coupling can take place.

In Figure 97, a leak to earth has developed from the 'live' conductor of a 100 V loudspeaker feeder. The other output terminal is connected internally to the amplifier case. This in turn is connected to the case of the mixer via the braid of the screened mixer lead. A resistor is shown here to indicate its impedance. The mixer case is connected to earth.

There is thus a complete circuit across the 100 V amplifier output via the earths and the braid of the mixer lead, and high-level signal currents flow. Owing to the impedance of the braid a voltage appears across it, and as it is carrying the input signal to the amplifier, this voltage is in series with it. The braid is thus the impedance that is common to both circuits.

The situation is similar to that of the earth loop, but in that case the 'dead' leg of the feeder was leaking to earth, thus producing a loop in which hum currents could circulate but no signal feedback. Here, the 'live' side of the feeder is earthed, so feeding the output signal into the loop.

There can be many forms of common-impedance coupling, both in the system wiring and in the amplifier itself, but earth leaks are a common cause external to the amplifier.

Dealing with instability

Like the hum loop, the cause of instability can be difficult to trace because so much of the system is affected by it. In the above case, all amplifiers will show spurious output readings and also the mixer because it too is included in the feedback loop. This gives the panic-inducing impression that the system has gone beserk and everything has gone wrong at once! However, there is usually just one cause and when it has been removed, all the meters will read normally.

When there are several amplifiers, quickly zero the gain control of each in turn, returning it if there is no difference. One should thus be identified as responsible. Next unplug each loudspeaker feeder from that amplifier in turn, until removing one clears the fault.

This leaves some section of the audience without sound. To trace the leak in a long feeder possibly difficult to access could take a while, but it may be possible to restore sound by a temporary expedient. With some amplifiers it is possible to change the output from unbalanced to balanced or floating by removing or repositioning links. If the output is converted to floating it is isolated from earth and the feedback loop is broken. If this facility is not available, a 1:1 isolating transformer to supply the affected feeder will have the same effect, if one is to hand.

Loudspeaker faults

These are quite rare but sometimes happen. Burnt-out coils which sometimes occur with high-powered hi-fi systems are extremely unlikely because there are usually so many units sharing the power that each dissipates a little, so most are liberally rated. For example, a LISCA system having 28 units in a medium-sized hall would have a power-handling capacity of 84 W if each was rated at 3 W. Yet the average power actually used would probably be no more than 10 W, 0.35 W per unit.

Coils can go open-circuit, the fault usually occurring at the soldered joint connecting the flexible lead-out wire and the coil-end. The joints are glued to the cone so care must be taken not to damage the cone when resoldering. The flexible wire must never be tight but have a certain amount of slack.

Small tears or holes in the cone can be repaired by sticking a small patch of stiff paper over them. Major damage, especially at the surround, cannot usually be successfully repaired.

Distortion, or more accurately, grating noises are sometimes produced, which are more noticeable at low volume. These are due to the coil going off-centre relative to the magnet pole-pieces. Old loudspeakers had centring spiders which could be adjusted with the aid of feelers inserted in the gap between the coil and the centre pole-piece. Now all units have domes which make the gap inaccessible, as well as glued centring rings, so centring is not usually possible. Grating can also be due to a distorted coil former or loose windings.

The fault can be identified by pushing the cone gently inward with the forefingers, whereupon the friction can be felt.

Most buzzes and rattles are due to the cabinet and particularly its back, so it is necessary to first eliminate the loudspeaker unit as the possible cause by means of the above test. It can be surprising just how full of buzzes and rattles a cabinet can be, whereas these pass unnoticed in normal use with only an occasional buzz on certain sounds.

To test for these, connect the loudspeaker to an amplifier fed by an audio oscillator. Turn the amplifier gain up fairly high, then sweep the oscillator from around 40 Hz upward. The buzzes will occur at certain frequencies and the offending cabinet part identified by feeling and pressing with the fingers. Often an extra screw at the critical point is all that is required to silence it.

Amplifier faults

Some experience and knowledge is required for the repair of modern electronic circuits. It is not advisable to attempt a repair without such experience as more harm than good may result. However, given some acquaintance with electronic circuitry, a successful repair can often be made by a non-professional. Skill in soldering printed circuits is vital as is the use of a multi-range meter.

A service manual for the particular model is almost an essential The circuits of old valve public-address amplifiers were predictable and varied little. Fault diagnosis was straightforward. This is not so with modern transistor amplifiers as there is a wide variety of circuit configurations in current use. The use of integrated circuits by many makers does not help as without the relevant information they are just a 'black box' with rows of unidentifiable terminal legs. It is a wise course, then to obtain a service manual or at least a circuit diagram of all equipment used in advance, and not wait for a fault to appear.

Many faults are revealed by taking voltage measurements and this should always be done as a first step, especially when the equipment is completely dead. It is assumed that the fuses have already been checked. If one has blown it is of no use to fit another and watch that one blow as well. On the other hand time can be wasted checking components when the fuse merely succumbed to an abnormally high surge.

A quick check often used by engineers is to connect an ammeter across the empty fuse holder and switch on. If the current exceeds the fuse rating, switch off immediately and start checking for the cause. If it is around half the rating it is probably about right. Even so, leave the meter in circuit for a while and keep it under observation in case the current starts to creep up due to the onset of thermal runaway or some other cause.

Often the condition of the blown fuse can give a clue as to the cause of its demise. A 'hard' blow in which the wire vaporized or left small globules of metal indicates a catastrophic cause, possibly short circuit output transistors or reservoir capacitor. A 'soft' blow where the wire has parted in the middle could be due to a surge, or a smaller excess current.

Voltage should be checked at the output of the rectifiers, not overlooking the negative rail if there is one. If there is no voltage, the a.c. from the mains transformer should be checked and, if this is absent, the transformer itself. Low voltage could be caused by an open-circuit rectifier diode, a low capacitance reservoir capacitor or excess current through the output transistors. Any of these will also cause hum.

Voltage regulators are often used for the earlier stages as they greatly reduce ripple on the supply, hence hum. They can go open-circuit, so voltage should be checked on the load side. If low, excess current could be the cause due to a fault in the circuits being supplied.

The output stage

Excess current in the output stage can usually be quickly detected by examination of the emitter resistors. These are generally 0.5–3 Ω and are of small wattage rating because their low value produces only a small voltage drop over them. They are thus the first to show stress if the current becomes excessive. They then become blackened and sometimes burn out. Any such indications point immediately to excess output current, and possibly a short-circuited output transistor.

The first thing to check in such a case is the loudspeaker system. A short-circuit or partial short-circuit reduces the load to below the minimum, produces high current and the likely demise of the output transistors. Repairing the amplifier and reconnecting into the system will only do the same thing again unless the loudspeaker fault has been put right.

If the loudspeaker circuit is in order, the fault could be an unprovoked failure of one of the output transistors. However, this cannot be certainly assumed, as with directly coupled circuits a defective transistor in an early stage can change the bias conditions throughout and produce excessive current in the output stage. A replaced output transistor could quickly be destroyed if such a fault remained.

Although it is quite a chore, the only certain way of avoiding this is to test each transistor right back to the input, or back to a capacitor coupled stage if there is one. In most cases this means removing them from the printed board and refitting them one at a time after testing.

Even with other faults, this is often the most practical way of tackling them, because stage-by-stage voltage readings, which constitute the standard method of diagnosis with capacitor-coupled stages, are often meaningless with direct coupling. One fault can change voltages everywhere. However, before doing this with a no-signal fault, a check along the supply rail and its feed resistors with a voltmeter may reveal an absence of voltage, perhaps due to a print break or short-circuit decoupling capacitor.

In comparing any voltage readings with those given in the manual, do not expect them to be exactly the same, as component tolerances will give deviations. Voltages should be close when measured from the output of regulators, and also voltage differences across base–emitter junctions.

Another cause of excess output transistor current could be maladjusted preset control. If fitted, this is usually connected between the two base circuits of the output pair, and it controls their bias. Current is measured by removing a link and inserting an ammeter, then the preset is adjusted to give the current stipulated in the manual. Wait for the output transistors to warm up before making the final adjustment.

The control preset could itself be the cause of incorrect current if damaged or there is tarnish on the sliding contacts. If the current changes in jerks and jumps when the control is rotated it should be cleaned or replaced.

When changing an output transistor, note that most have their collectors internally connected to the case. As they are bolted to a heat sink which is at chassis potential, they are electrically insulated by a mica washer that conducts heat. Care must be taken not to damage this washer when removing the old transistor and to ensure it is not omitted when the new one is fitted. Insulating bushes are also used for the fixing screws with this type of transistor, and these too must be refitted. The nuts must be done up tight to achieve good thermal contact, and thermal conductivity is often aided by means of white zinc oxide paste.

If one or both of the output transistors have become short-circuited due to a low impedance load, with some circuits it is likely that the driver transistor has also suffered. The reason is that when the output transistors are short-circuited, with direct coupling the driver is effectively connected to the load, so it suffers the same fate as the output transistors. Examination of its emitter resistor will usually indicate whether this is so.

Capacitor-coupled stages

Some public address amplifiers use capacitor-coupled stages for part if not all of their circuits. Faults in these are easy to diagnose by simple voltage measurements. A deviation greater than about 10% from those given in the manual should be investigated as probably pointing to the fault.

If no manual is to hand, it is still possible to trace a fault by checking voltages and comparing them with those normal for that type of circuit. A basic class A capacitor-coupled stage with expected normal voltages is shown in Figure 98.

Voltage at the collector is generally half that of the supply. This enables equal voltage swings either side of this value to follow the positive and negative peaks of the input signal. For low-signal circuits such as input stages, the danger of clipping one half cycle through overloading is remote, so the collector resistor

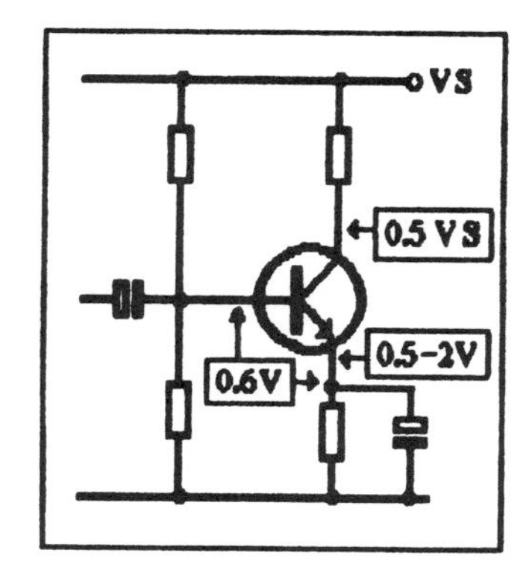

Figure 98 Typical voltages for capacitance-coupled stage.

may be of a higher value to achieve high gain. In this case the collector voltage would be less than half the supply value. Current drawn by the meter through the resistor would make the measured reading lower still.

High collector voltage and no voltage on the emitter indicates that the transistor is not passing current and may be open-circuit. The same effect would be obtained if the base resistor were open-circuit, as there would then be no forward bias (Figure 99).

When the collector voltage V_c is low and emitter voltage V_e high the cause could be a leak in the transistor, an open-circuit lower bias resistor, or a leak in the base coupling capacitor (Figure 100). The last two result in excessive forward base bias. It may be noted that the lower base resistor is now seldom used. A common configuration is to return the supply base resistor to the collector, thus achieving negative feedback and low noise, and dispensing with the lower one. In such a case the value of the base resistor is higher than in the circuit shown.

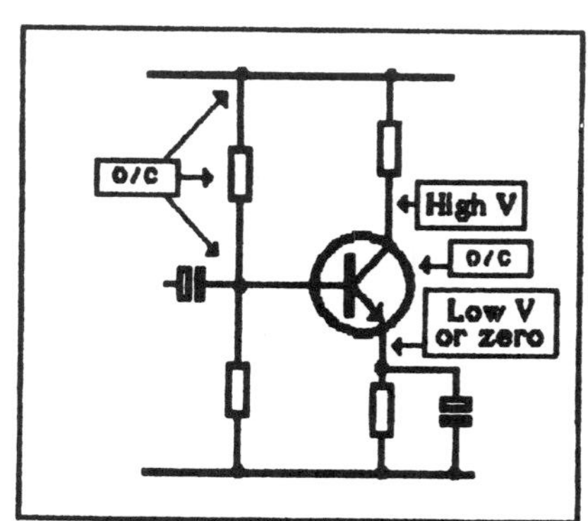

Figure 99 Possible faults when V_C is high and V_e is low.

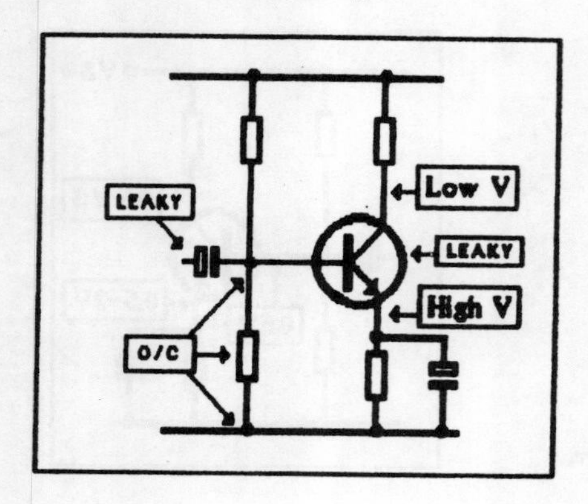

Figure 100 Possible faults when V_c is low and V_e is high.

The emitter bypass capacitor serves to prevent negative feedback over the emitter resistor, which would considerably reduce the stage gain. Sometimes the emitter resistor is split with part left unbypassed to achieve some negative feedback. In cases of low gain, if voltages seem about right, an open-circuit emitter capacitor is a likely cause.

Amplifier hum

The reservoir capacitor is a prime suspect due to loss capacitance, but check also that the can is not loose but securely fixed in its clip to the chassis. Faulty capacitors often have a bulging end-seal or white powder appearing around the terminals, but not always, the capacitor can be faulty and look perfectly sound.

The value is usually too high for direct reading with most capacitance meters, so the only test is to disconnect and replace with one of similar value. Bridging it with another is not a valid test as the extra capacitance may reduce hum that has its cause elsewhere. When replacing a reservoir capacitor make sure that the voltage and ripple current rating of the replacement is high enough. The latter is sometimes disregarded, leading to premature failure. It should be at least as high as the total supply current at full power.

Another possible cause of hum which could be overlooked is excessive current, usually through the output stage. It may not be sufficient to cause thermal runaway and destroy the transistors, and so may not be suspected. The discharge slope of the reservoir capacitor between successive charging half-cycles depends on the discharge current. The greater the current the steeper the slope and the greater the ripple. (See Figure 53, Chapter 7).

So ripple is increased by excessive current, but it can be reduced by increasing the reservoir capacitance. This is why bridging the reservoir capacitor with another may reduce the hum, but merely masks the real fault. A check should therefore be made on the output current, and an adjustment made if required, before condemning the reservoir.

Open or short-circuited diodes in the bridge rectifier can cause hum, and these can be easily checked by a resistance measurement. With full-wave circuits another possibility is that one half of the mains transformer secondary is open-circuit, although this is not common.

Voltage regulators are often used to smooth the supply to earlier stages that are more vulnerable to supply ripple because of high following gain. A short-circuit regulator would have little general effect but would produce hum. Measuring the voltage on either side will reveal if this is so. If it is the same, the device may be short-circuited, but if the load side is lower, and is the specified output voltage, it is in order.

Another possibility is that the supply voltage is too low. If the input voltage to a regulator is lower than its specified output voltage, it does not regulate, but just passes current through. Input and output voltages are therefore the same, but this is not the fault of the regulator but the supply.

Hum can be caused by hum loops within the amplifier, but these would be the result of poor layout design and not suddenly occur in an amplifier that was previously hum free.

Amplifier noise

This may consist of either continuous high-level hiss or crackling and tearing sounds. In either case the cause can be partly localized by operation of the volume, gain, or fader controls. If a control stops it then most likely it is in a stage prior to the control. If it has no effect, it is virtually certain to be in a later stage.

This test is not entirely conclusive as a component breaking down, especially in the power supply circuits, may create noise impulses that may be picked up by earlier stages. This may well be the case if a control reduces the noise but does not completely eliminate it at its lowest setting.

Capacitors are the most likely cause of tearing and crackling. A signal tracer consisting of a small amplifier and loudspeaker with an input probe may help to locate the trouble. Another simpler method, though not quite so reliable, is to short various points of successive stages to chassis with a high-value capacitor on a lead (1–10 μF should be suitable). The fault is most likely to be before those points where the capacitor stopped the noise and after those at which it did not.

Tracking across adjacent print runs is another possibility. It happens when conductors at high potential are close to those that are low, and is started by some conductive dirt or deposit between them. Once started, heat is generated, which chars the intervening board, and the carbon so produced conducts still more.

Although common in some types of equipment such as television receivers in which high potentials exist, tracking across print runs is less so with amplifiers which have a typical supply potential of no more than 100 V. However, if an elusive noise cannot be traced it is worth making an examination of the printed board, and further investigating any darkened portions.

If such is discovered, the only remedy is to clean the affected area thoroughly and chip away all the charred material even if this produces a hole in the board. Smearing on insulating paste such as silicone grease is useless as the tracking will continue underneath it.

Steady hiss is nearly always associated with a defective transistor, although low-wattage resistors also generate noise which can increase. The gain-control test is more conclusive here, but remember that all transistors generate noise, and hiss will be heard if all controls are turned fully up with no input. The signal tracer will help to locate any source of excessive hiss. If there also seems to be abnormally high gain, a negative feedback fault would be indicated, although this is less likely.

Amplifier distortion

A common cause of distortion is low output-stage current producing Class B cross-over effect. Adjust to the correct value if a preset adjustment is provided, if not check the biasing circuits and voltages at previous stages if directly coupled. The driver stage could have a fault which caused it to be overloaded by the high input signal present there.

Distortion with low output is nearly always due to a defective output stage being overloaded in the attempt to get more gain. A faulty output transistor is likely especially if one of a complementary pair. Surprisingly, one transistor of a pair driving a transformer can sometimes be completely open-circuit and have little audible effect with a moderate load. It is when the amplifier is fully loaded that distortion occurs due to overloading the single good transistor.

A less common but not unknown cause of low output with distortion is short-circuited turns in the output transformer.

Resistance measurements of the winding does not help, because the resistance difference caused by a single short-circuited turn or even a small adjacent group is negligible. As transformers are expensive, substitution is obviously to be avoided unless diagnosis is certain. One method of testing is by using an impedance tester. Disconnect the primary connections and apply the tester. It should show a high impedance. If in doubt connect a 4 Ω loudspeaker across the 4 Ω tap, whereupon the impedance should drop considerably. If there is only a small change, the transformer is already heavily loaded by short-circuited turns.

A further cause of distortion is similar to that described for external faults, which is supersonic instability. A steady reading on the output meter with no input indicates this.

Amplifier instability

Distortion caused by supersonic oscillation or audible whistles is the symptom of instability. The first check is to ensure that the cause is not external by disconnecting everything except a monitor loudspeaker.

The cause is nearly always common-impedance coupling through the supply circuits due to the failure of a supply decoupling capacitor. Diagnosis is by simply bridging each decoupler in turn with a good one until the oscillation ceases.

Instability or hum after a repair could be due to some alteration to the wiring, especially the earthing. In particular the screens of all screened wires must be earthed, but *only at one end*. Any new earth wiring should be to the same point as the original.

Poor sensitivity

If confined to one particular input, poor sensitivity will clearly be due to the associated input circuit. If all are affected the trouble is in a later stage. An open-circuit emitter bypass capacitor is a likely cause. Low supply voltage, possibly due to a leaky decoupling capacitor causing excessive voltage drop over a series resistor, is another. Resistors do not often change value except high-value ones above 1 M Ω. These sometimes go higher or completely open-circuit—it is worth checking their value.

Using the oscilloscope

The oscilloscope can be a powerful tool in diagnosing amplifiers faults. Its use as a mixer monitor has already been mentioned. Low ambient noise can be observed, indicating a live microphone; non-linearity in the mixer can be detected as a difference in the average amplitude between half-waves above the zero line to those below. The onset of clipping can be seen as a gain control is advanced, before distortion becomes audible.

When checking signals in an amplifier or from signal sources and in other equipment, faults and various conditions can be readily identified. Here are a few:

Phase differences between two channels, between an input or output or between two amplifiers can be observed by feeding in a sine wave and applying the output of one channel to the vertical and the other to the horizontal scope input. If in phase, the displayed figure is that of two parallel lines tilted at 45° with the left-hand down. As the phase difference increases the lines part to form an ellipse, a full circle is obtained at 90°. Beyond this the circle narrows to an ellipse tilted in the opposite direction until it again becomes a pair of parallel lines at 180° (Figure 101).

Clipping appears as a flat top to a peak, which indicates that a stage is being overloaded by too high an input signal. When both halves of a sine wave are clipped equally, the clipping is said to be symmetrical and shows that the affected stage is operating in class A, or if it is the output stage, that the two sections are balanced. If one half-wave clips before the other as the gain is increased, it is asymmetrical, and indicates that the stage is not operating at the centre of its transfer curve, or that the output stage is unbalanced (Figure 102).

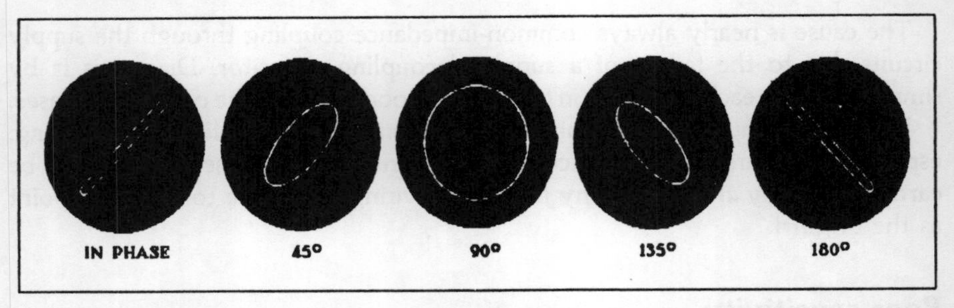

Figure 101 Oscilloscope traces showing phase relationship of two signals, one applied to horizontal (X) input and the other to the vertical (Y).

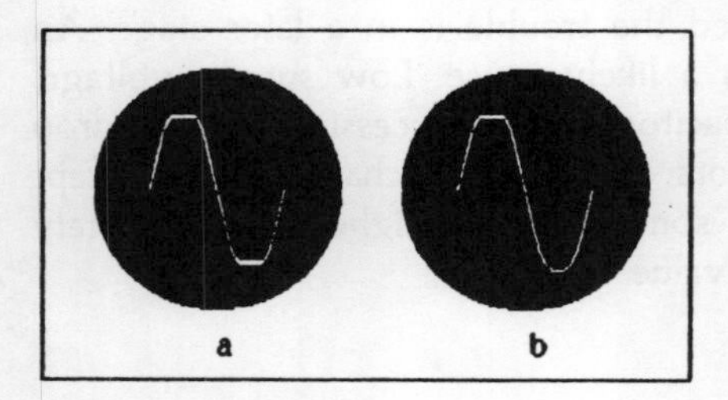

Figure 102 Clipping: (a) symmetrical; (b) asymmetrical.

The power rating of an amplifier can be checked by fitting a dummy load of the correct impedance and adequate wattage to the output terminals, along with a voltmeter and oscilloscope. An audio oscillator is connected to the input and a sine wave fed in. The input amplitude is increased until clipping appears on the waveform shown on the scope, then reduced slightly until the clipping just disappears. The output voltage is noted and the wattage calculated from $W = E^2/Z$, where Z is the output impedance, and E is the voltage (Figure 103).

Distortion is often due to clipping, but there are other types. Harmonic distortion occurs due to various non-linearities in the circuits. Incorrect bias is a common

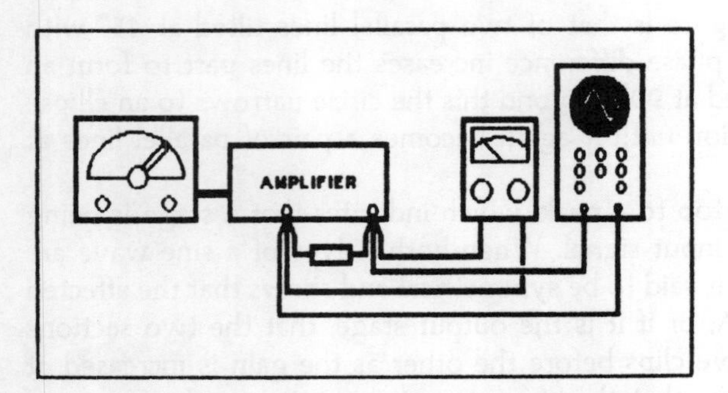

Figure 103 Equipment set-up for power testing an amplifier.

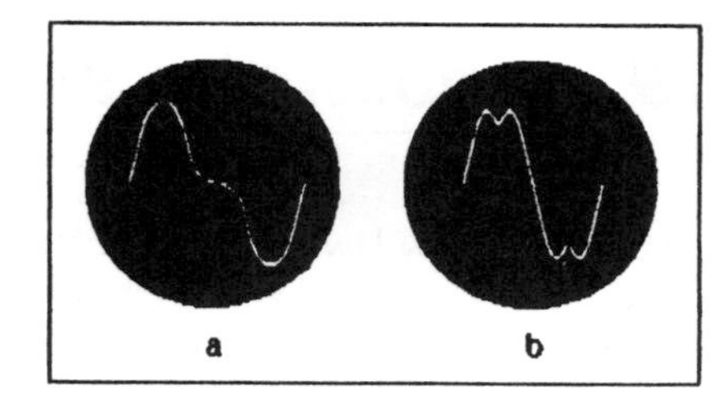

Figure 104 Harmonic distortion: (a) even; (b) odd.

cause. Spurious harmonics are thereby added to the fundamental waveform so distorting its shape. These can often give a clue as to the cause. Even and odd harmonic distortion are shown in Figure 104.

Frequency response is of concern with mixers that have tone controls. Their operation, in particular whether the centre point of a plus and minus control gives a flat response, can be quickly checked by the scope and a square-wave generator set to about 1 kHz (Figure 105).

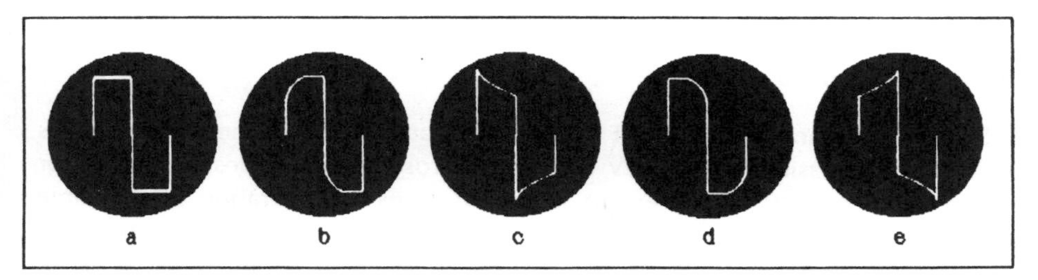

Figure 105 Determining frequency response from square wave: (a) ideal square wave; (b) poor h.f.; (c) excessive h.f.; (d) poor l.f.; (e) excessive l.f.

A flat response produces an undistorted square wave (a). Rounded leading edges indicates a falling high-frequency response (b), while raised leading edges denote a boosted high-frequency response (c). Rounded trailing edges are obtained with reduced low-frequency response (d), but raised trailing edges show boosted low-frequency response.

The easy way to remember these is to think of the leading edge as governing the treble and the trailing edge the bass. In both cases, a low edge denotes a low response while a raised edge means a raised response.

Incipient instability This is a useful test as it identifies an amplifier that is on the point of instability or likely to become unstable. Some amplifiers become unstable if there is excessive capacitance in the output circuit. With no input and the scope connected to the output with no load, various values of capacitor from 0.01 to 0.1 μF are connected across the output terminals. No spurious waveform should appear.

Another test is to feed in a square wave. Any waviness along the top denotes incipient instability which has been triggered by the sharp rise of the square wave.

19 Appendix: safety regulations

A number of new laws and regulations have come into force to supplement the Health and Safety at Work Act of 1974. These are mostly embodied in the Electricity at Work Act, which came into force in 1990, and guidelines issued by the Health and Safety Executive (GS 50, 1991). Licences for public gatherings can be refused by local health and safety inspectors if they consider the requirements have not been met, but their interpretations can vary.

As affecting public address systems, the rules concern mains supplies, earthing, 100 V feeders, and cable runs. Some have been noted in the main text of this book, but there are also practices which, although they are technically sound and have been used safely for years, are now deemed unacceptable. Suggestions made in previous chapters should therefore be viewed in the light of the latest regulations and may need to be modified accordingly.

Mains supplies

All supplies to any system must be from the same phase of the mains. This is to avoid the possibility of 400 V appearing across insulation designed only for 240 V. This is easy to implement where all items are operating in the same location, but could cause a problem of phase identification where the mixer is remote. For permanent installations this could be checked out, but it may be difficult with a temporary, large installation such as a sports stadium.

A supply feed from the main control room is one possibility, but earthed flexible armoured cable is preferred and may be demanded by the inspectorate. This could be expensive and inconvenient. Another solution is to use a battery operated mixer. Mixers generally require low supply current if not overly elaborate, so it may be possible to convert a mains-powered mixer to run on a battery. A means of checking the battery in situ should be devised, perhaps by switching the output meter to read voltage. A spare battery should always be available of course.

Residual current devices (RCDs) rated to trip at 30 mA should be included in all mains supplies. Ideally these should be on the sub-circuit distribution board and each RCD should supply no more than 6 outlets. Alternatively they could be incorporated in individual sockets but these are less economical. RCD adaptors or plugs should not be used as they can be easily bypassed or removed.

For permanent installations the RCD must be tested by operating the trip button at least once a month, but for temporary systems it must be tested daily before the start of the event.

Protection against excess current must be provided by the usual fuses of correct rating for the equipment or by circuit breakers (CBs). These may be combined with residual current devices (RCBOs).